Handboo
in Rodents and Rabbits

MW01416023

Kathleen R. Pritchett-Corning, DVM, DACLAM, MRCVS
Director, Research and Professional Services
Charles River Research Models and Services

Aurélie Girod, DVM, Dip. Vet. LAS.
Manager, Professional and Veterinary Services
Charles River France

Gloria Avellaneda, DVM
Scientist
Charles River Avian Products and Services

Patricia E. Fritz, VMD, MS, DACLAM
Principal Investigator
Consulting and Staffing Services
Charles River Frederick

Sonja Chou, VMD, MS, DACLAM
Director, Veterinary Services
Charles River Preclinical Services

Marilyn J. Brown, DVM, MS, DACLAM, DECLAM
Executive Director, Animal Welfare and Training
Charles River Research Models and Services

Published by Charles River Laboratories ©2011
2nd Edition - April 2011

Table of Contents

As a person who works directly with animals, your job is one of the most important in biomedical research. Veterinarians, researchers, and the animals themselves rely on you to notice changes and help ensure the well-being of the animals in your care. Recognizing clinical signs of illness or stress is not only a welfare issue, but a scientific one, as some of these conditions have the potential to disrupt research projects.

This book is organized into sections, and is written as an introductory guide. The first section is the most important as it will teach you how to observe the animals you work with and how to report clinical signs. The other sections describe, first in words, then in pictures, some signs of illness, stress, or injury you may see when working with rodents or rabbits. These sections first address the whole animal, then move from nose to tail for specific observations related to various parts of the animal.

We hope you find this book a useful reference and that it enhances the care of the animals with which you work. If there is a clinical sign you feel the book is lacking, please submit comments or suggestions to askcharlesriver@crl.com.

The Veterinary Staff of Charles River

Disclaimer

Acknowledgements

Editorial assistance from Dr. Charles Clifford, Dr. Guy Mulder, Dr. William White, Marie-Odile Bideau, Mort Alling, Sarah Warren, and Jason Ocker. Design assistance from Pamela Clasby and Robert Zaccardi. Special thanks for the use of photographs to Dr. Paul Flecknell, LAVA UK, Dr. Joseph Garner, Dr. William White, Dr. Takashi Kuramoto, Dr. Jennifer Kalishman, and Dr. Sonya Gearhart.

Watching the animals in your care for changes in their behavior or bodies is an important part of your job. It is something you should feel comfortable doing. To notice a change in an animal, however, you must first know what is normal. During the usual course of your work, there are many opportunities to notice the usual appearance and behavior of animals. What is normal for an animal or particular group or type of animal should always be held in mind when making an observation. When we observe a change from normal in an animal, this change is called a **clinical sign** because it could be a sign of a clinical (or visible) disease. Clinical signs may include things like sneezing, diarrhea, or a change in appetite. We may also call this change from normal an **abnormal finding** — something that you found on or about the animal that was not usual. This could, for example, include a cut or a scrape, a lump, or a change in behavior. There is no real difference between clinical signs and abnormal findings—they are just different ways to refer to changes from normal in an animal. This book does not describe every possible illness, injury, or abnormality in rodents and rabbits. Its goal is to illustrate some of the more common clinical signs and to teach you how to make a clinical observation.

It is not difficult to make observations. You need to be alert to changes in your animals, no matter how small the animal or the change. You have an advantage because you know the animals you work with very well and you may be able to spot small changes that others might not see. To make these observations, you need to use your sight, your hearing, your sense of touch, and maybe even your sense of smell. Practice making an observation with the exercise on the next page.

What is this?

1)   An orange.

2)   A round or spherical orange-colored object. Its surface appears dimpled and areas of the surface are shiny. There is an indentation at the top, which contains a small bit of green material.

Answer #1 is a **diagnosis**. It is the conclusion that you reached after looking at the picture and comparing it to your mental description and experiences of oranges. Answer #2 is an **observation**. It is what your senses tell you and what you can measure or objectively test. After all the data are gathered by observation, you can say, "This is probably an orange." Since we made this observation on a photo, we could not evaluate things such as weight, smell, or sound. Those are some other measures or objective evaluations that might be made by an observer.

When making an observation, be sure to look at all aspects of the animal. With small, caged rodents it is easy to open the cage and look only at their backs. Be sure to examine both the whole animal and its housing—nose to tail, top and bottom, excretions and environment. This examination may provide you with valuable clues to the problem and will give the veterinary or research staff the entire picture.

If an abnormal animal is seen on a quick viewing of the room, you should follow a certain series of actions in order to perform a thorough examination and provide information to the veterinary or research staff. If the same steps are followed every time, nothing will be missed and the quality of your observation will be better. Practice makes this much easier; eventually it will be automatic. An example of a list of actions is below.

Clinical Observation Flow if an Abnormal or Ill Animal is Seen
1. Examine the cage from the outside and note the conditions in the room.
2. Open the cage.
3. Look at the water and feed supply.
4. Look at the walls of the cage and the bedding.
5. Observe the behavior and attitude of the animal in the cage.
6. Pick up the animal.
7. Look at the back of the animal.
8. Look at the belly of the animal.
9. Look at the extremities of the animal (tail, head, limbs, paws).

When making notes on an observation on animals in your care, you need certain information to give to the veterinary staff or researcher. Reminders of what information is required may be provided for you in a clinical examination form.

Parts of a Clinical Observation
1. The **species, strain, age or weight, sex,** and **history** (if known) of the animal
2. The **observation** itself
3. **Where**, on the animal, the abnormality was found
4. The **size** of the abnormal finding or a **description** of the clinical sign
5. **Duration** the abnormality has been present (if known)
6. Whether or not **other animals have similar signs**
7. **Any other description** that might help make a clear picture for someone unable to see the animal

For example, if you were to notice that some mice in a cage of haired mice had hairless areas on their bodies, your clinical observation might look like this:

A Clinical Observation
1. Mice, C57BL/6, six-week-old females, on experiment for Dr. W.
2. Hair loss; skin underneath seems healthy
3. Hair loss seen on the sides and back
4. Areas of hair loss are half-inch in diameter
5. First noticed three days ago and has worsened since
6. Four of five animals in cage affected
7. No other abnormal findings; no hair seen in cage

This observation includes the vital information described in the first table. Measurements do not have to be exact. If you have no way to measure something, approximate as closely as you can using common objects that most people will understand.

As you make observations of the animals in your care, you will be working with facility management and the veterinary staff. They will build on the observations you make, maybe by asking questions about the environment, or perhaps by asking that you send the animal, its cagemates, or other animals from the room for diagnostic testing. Your veterinary or research staff may keep you informed so that you know what diagnoses were made from your observations. We have included as **Appendix A** a short glossary of some medical terms you may encounter, and also diagrams illustrating veterinary terminology for locations on the body (Figures A, B, and C). As you will see, several different diseases could cause many of the clinical signs that you might observe. Further testing of blood or tissues is often needed to determine which disease is causing the signs you see. For example, a lump on an animal may be cancer. On the next animal with a lump, that lump might be caused by an infection. Only specialized examination of the lump will show the difference between the two.

Remember, you will never be wrong if you describe carefully, exactly, and completely what you see, hear, and smell about the animal and its environment.

Descriptive Terms

Orientation and Regions with Respect to the Body

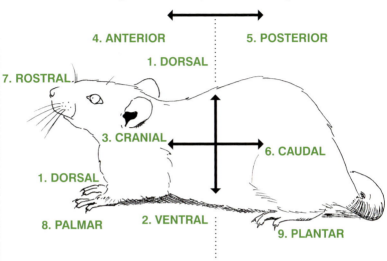

4. ANTERIOR   5. POSTERIOR

1. DORSAL

7. ROSTRAL

3. CRANIAL   6. CAUDAL

1. DORSAL

8. PALMAR   2. VENTRAL   9. PLANTAR

Figure A

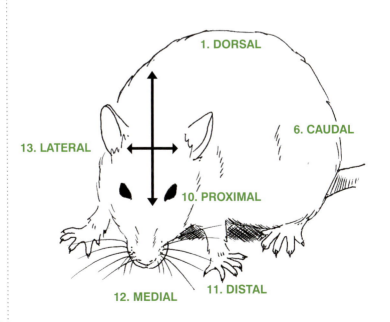

1. DORSAL

6. CAUDAL

13. LATERAL

10. PROXIMAL

12. MEDIAL   11. DISTAL

Figure B

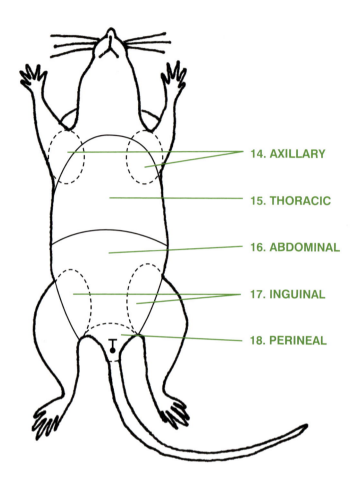

14. AXILLARY

15. THORACIC

16. ABDOMINAL

17. INGUINAL

18. PERINEAL

Figure C

10

1. **DORSAL**: The surface directed towards the back, spine or the top surface of the limbs.

2. **VENTRAL**: The surface directed towards the belly or the ground.

3. **CRANIAL**: Towards the head end of the body.

4. **ANTERIOR**: The front half of the body.

5. **POSTERIOR**: The back half of the body.

6. **CAUDAL**: Towards the tail end of the body.

7. **ROSTRAL**: Towards the head or the mouth.

8. **PALMAR**: The surface below the proximal ends of the wrist, directed towards the ground.

9. **PLANTAR**: The surface below the proximal end of the ankle directed towards the ground.

10. **PROXIMAL**: Closer to the long axis of the body.

11. **DISTAL**: Distant from the long axis of the body.

12. **MEDIAL**: The surface towards the median plane of the body.

13. **LATERAL**: The surface directed away from the median plane.

14. **AXILLARY**: The area pertaining to the axilla (armpit).

15. **THORACIC**: The area pertaining to the thorax (chest).

16. **ABDOMINAL**: The area pertaining to the abdomen.

17. **INGUINAL**: The area pertaining to the groin.

18. **PERINEAL**: The area pertaining to the region between and around the anus and genital organs.

Figure 1 These photos also illustrate normal healthy animals. Normal animals appear as they should for their strain, age, and sex. They are neither thin nor fat (unless, like some strains of rats and mice, they are supposed to be). They have well-groomed haircoats, and no discharge from their ears, eyes, mouth, or nose. Their fecal and urine output is neither too much nor too little and is of an expected frequency, color, and consistency. Their breathing is quiet and regular, and they breathe through their noses, with closed mouths. If they are awake, they are alert and notice changes in their surroundings. They interact with their cagemates or handlers, and they move freely and comfortably around the cage. Other features of normal animals are illustrated in **Figure 2**.

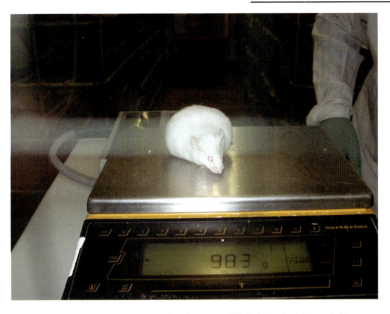

Figure 2A A late-term pregnant outbred mouse (CD1). Note that her weight approaches 100 g.

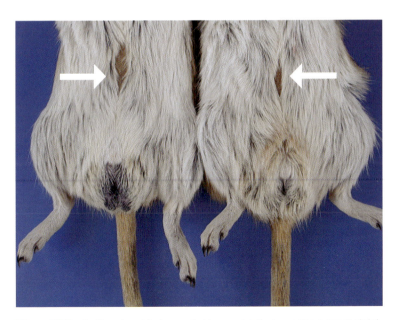

Figure 2B The bellies of gerbils have a hairless patch that contains a scent gland (arrows). This scent gland is larger in males (on the left).

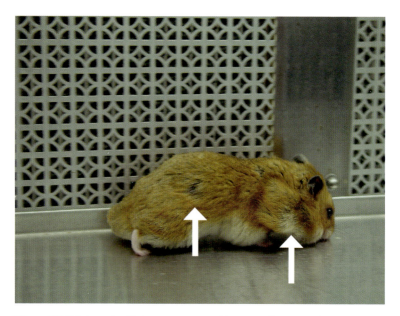

Figure 2C This hamster illustrates two normal features of hamsters (arrows); the cheek pouch (stuffed with food) and the flank gland, a scent gland more prominent in males.

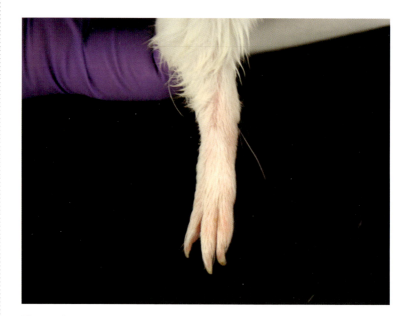

Figure 2D A normal guinea pig hind foot has three toes.

Body condition scoring
Figure 3

Description: A means of providing an objective measure of how fat or thin an animal is. This measure is accomplished through observation or palpation of the animal, and then the physical findings are compared to a predetermined scoring system. A numerical score is assigned to the animal, and then recorded.

Reasons: Body condition scoring can be a useful animal welfare tool. It allows for the determination of endpoints in studies where animals are expected to lose or gain weight. For some studies, determination of endpoints may result in better welfare for animals.

Thin (Emaciated, cachectic, anorexic, skinny, underweight, underconditioned)
Figure 4

Description: An animal is thin when it has a lower body weight than expected for its skeletal size. Animals may be thin but have swollen bellies. If you handle a thin animal, it may feel bony or delicate.

Reasons: Animals become thin when the amount of food they are able to take in is less than the amount they need. This condition may be due to many causes, including infectious disease; problems with the mouth, teeth, or tongue; the demands of a nursing litter; or a high metabolic rate.

Fat (Obese, overconditioned, overweight)
Figure 5

Description: An animal is fat when it has a higher body weight than expected for its skeletal size.

Reasons: Animals become fat when they cannot use up or burn off the amount of food they take in. Excess calories are stored in the fatty tissue. In some cases, fatness is due to a genetic defect and is an expected trait. Animals may also gain weight as they age, especially if they have unlimited access to food.

Undersized (Dwarf, runted)
Figures 6, 33A

Description: An animal that is smaller than normal for its Determination of endpoints age and sex.

Reasons: Animals may be undersized due to genetic mutations that make them smaller overall. This condition is rare, but some strains of mice are bred for it. The most common reason for being undersized is runting (i.e., being the "runt of the litter"). This animal is usually small, weak, and appears sickly. It may have had too much competition in the uterus, or it may have been weak for some other reason and did not get enough milk. (See *hydrocephalus, malocclusion*)

Enlarged abdomen
Figure 7

Description: An animal whose belly is larger than normal for its age, sex, breeding status or strain.

Reasons: Animals may have enlarged abdomens when they retain fluid in their bellies. This condition can be due to a problem with their kidneys, liver, or heart, or certain types of cancers. Enlarged abdomens are normal in pregnant females. (See *lumps* or *swellings*)

Hunched (Depressed, kyphotic, painful, stretching)
Figure 8

Description: A hunched animal is one that has drawn its abdomen up and has its head down and its four feet underneath it. It may be quiet and still, or it may stretch upward and then settle down. The animal's eyes are often closed or half-closed.

Reasons: Animals that are ill for many different reasons may appear hunched. Animals may be hunched due to pain in their abdomens or other places. They may be hunched due to pain-causing infections or disease or because they are having difficulty breathing. Animals may also appear hunched when they have an upward deviation of their spine (kyphosis), but this condition is rare and almost never seen. (See *ruffled*)

Ruffled (Poor coat condition, ungroomed, piloerection)
Figures 8, 9

Description: A ruffled animal is one that shows an ungroomed hair coat in which the hairs are separated into little clumps that stand on end. It may also look greasy.

Reasons: Animals may be ruffled for many reasons. The animal's coat could be standing on end (piloerection) because the room is cold. The animal may be in pain or ill and therefore not devoting energy to grooming. The animal may be too fat to groom effectively. The animal may be of a stock or strain that tends to become greasy and ungroomed with age. (See *fur staining, hunched*)

Pale or blue (Anemic, cyanotic)
Figure 10

Description: Normally, hairless areas on an albino animal are pink or red due to the presence of blood circulating close to the surface. These areas include the ears, the inside of the mouth, the inside of the eyelids, and the tail and feet. This pink tinge may be seen to a lesser degree in animals with pigment. The eyes of an albino animal are dark red. A pale animal will appear white, blue, or a lighter shade of red in those areas.

Reasons: Animals are usually pale if they have a decrease in the number of red blood cells circulating in their blood, which is called anemia. This condition may be due to blood loss after surgery or from a wound, internal bleeding, kidney failure, cancer of the bone marrow, or some types of problems with the immune system. Animals may also appear pale if their blood vessels constrict due to cold or certain drugs, resulting in less blood flow to the animals' skin or mucous membranes. Animals that are blue may have defects in circulation or may not be taking in enough oxygen.

Yellow (Jaundiced, icteric)
Figure 11

Description: Normally, hairless areas on an albino animal are pink or red due to the presence of blood circulating close to the surface. These areas include the ears, the inside of the mouth, the inside of the eyelids and the tail and feet. This pink tinge may be seen to a lesser degree in animals with pigment. A yellow animal will appear pale or bright yellow in those areas.

Reasons: Animals become jaundiced when they have problems with the gall bladder or liver disease. When red blood cells break down, one of the byproducts of the breakdown and the body's metabolism of the dead cells is a yellow substance called bilirubin. This product is processed by the liver and excreted in the bile and feces. If this substance builds up in the blood or tissues, animals will appear yellow.

Masses or swellings (Lumps, bumps, tumors, abscesses, cysts)
Figure 12

Description: An animal has lumps or swellings on various parts of its body.

Reasons: Without a veterinarian opening the lump or swelling, it is difficult to tell what it might be. If the lump is located under the skin, it could be a tumor, an area of infection, or a fluid-filled structure like a cyst. A lump in the abdomen might be those things, but it might also be a normal pregnancy or a fetus retained inside the mother after it was to be born. Lumps and swellings might also be due to experimental procedures. (See *enlarged abdomen*)

Puffed or swollen mice ("Bubble mice," [subcutaneous emphysema], generalized fluid retention under the skin [anasarca])
Figure 13

Description: The skin of a mouse appears to be swollen away from its body. The condition usually occurs in nude mice and is rare.

Reasons: Nude mice are prone to cancer of their white blood cells and lymph nodes. This cancer may spread to their lungs and chest cavities. This condition causes problems with breathing. Occasionally, when they have this cancer, they damage themselves by breathing too hard and tearing their lungs. The rents in their lungs then allow air to escape into their bodies, where it migrates to their skin. This condition may also be seen if a gavage procedure damages the structures of the throat and chest. Generalized fluid retention under the skin is seen with severe kidney, liver, or heart disease. The swelling is more "droopy" than the firm puffing seen with air under the skin. It is also more generalized, whereas air under the skin tends to occur in pockets.

Abnormal breathing (Respiratory distress, gasping, dyspnea, wheezing, sneezing, snuffling)

Description: An animal has difficulty breathing. You may be able to hear them breathing or wheezing. Sometimes the noises may also sound like rattling, snicking, or snuffling. Animals with difficult breathing may also breathe more quickly than normal or show great effort when taking each breath. Animals may be thinner than normal, and also hunched and ruffled.

Reasons: Animals usually have difficulty breathing when something is wrong with their nose or lungs. This condition might be caused by an infection, a mass or fluid in their abdomen pressing on their chest cavity. The noises associated with breathing may be due to fluid in the lungs or a discharge in the nose or throat. Swelling of the nose or throat may also cause noisy or labored breathing.

Abdominal rigidity or writhing
Figure 14

Description: An animal may contract its belly abruptly, making it appear as if something is moving in its abdomen (writhing). Animals may also hold their bellies stiffly, resisting any attempt to move.

Reasons: Abdominal writhing or rigidity is seen when an animal has abdominal pain. It may also be seen when the animal is in pain from a different area of its body.

Wet-dog shakes

Description: An animal shakes its skin and fur as if it were wet and trying to dry itself. The shake usually starts at the nose and continues to the tail. It involves the whole skin, not just patches. This condition is not a tremor or a fine vibration of the animal

Reasons: Animals may shake if they are wet and trying to dry themselves. This condition may be part of the normal grooming pattern for that animal, as it tries to settle its fur after a long grooming session. Animals may shake their fur if they have itchy skin. Animals may also show wet-dog shakes after administration of certain compounds.

Loss of appetite (Anorexia, inappetance, not eating)

Description: An animal may not be eating a normal amount of feed.

Reasons: Anorexia is due to many potential causes. The animal may be in pain from its mouth or other areas. It may find the feed unpleasant. An animal may not have enough water to comfortably eat its feed, or it may be too weak or ill to eat. (See *thin*)

Abnormal posture (Splay leg, broken back, other gait or movement abnormalities)
Figure 15

Description: An animal assumes a posture that is not usual for the species, or it may appear uncomfortable, as if it cannot move from that posture.

Reasons: Animals may show abnormal postures if their backs or hips are injured or painful. Animals with spinal cord injuries will often show abnormal postures. Some abnormal postures may be related to damage to the brain or ears. (See *head tilt, abnormal movements*)

Weakness (Paresis)

Description: An animal lacks the strength to move or to hold itself in normal postures. Weakness can be of the entire body or of a limb.

Reasons: Weakness is a common finding in many disease states or injuries. It can be related to pain, injury, lack of food, lack of water, or exhaustion.

Dehydrated
Figure 16

Description: An animal has low body fluids, resulting in dry mouths and eyes. When an animal is dehydrated, its skin will remain "tented," or pulled upward, when the skin is gently pulled away from the body and then released.

Reasons: The most common cause of dehydration in laboratory rodents is lack of water (they cannot reach the sipper tube, the water bottle is empty, there is no pressure in the automatic watering system, etc.). Animals become dehydrated when they lose fluids or do not take as much fluid in as they lose. Animals can lose fluid through diarrhea, blood loss, or leakage of body fluids through open sores. Animals may also lose fluids through evaporation during surgery. Animals may not drink enough to replace normal fluid losses when they are in pain or weak due to other diseases. (See *nasal discharge, diarrhea*)

BC 1
Mouse is emaciated
- Skeletal structure extremely prominent; little or no flesh cover.
- Vertebrae distinctly segmented.

BC 2
Mouse is under-conditioned
- Segmentation of vertebral column evident.
- Dorsal pelvic bones are readily palpable.

BC 3
Mouse is well-conditioned
- Vertebrae and dorsal pelvis not prominent; palpable with slight pressure.

BC 4
Mouse is over-conditioned
- Spine is a continuous column.
- Vertebrae palpable only with firm pressure.

BC 5
Mouse is obese
- Mouse is smooth and bulky.
- Bone structure disappears under flesh and subcutaneous fat.

BC=body condition

Figure 3A Mouse body condition scoring. Body condition scoring is a way to assess the overall health and weight of an animal through sight and palpation.

Redrawn from Ullman-Cullere, M. H. & Foltz, J. Body condition scoring: a rapid and accurate method for assessing health status in mice. *Lab. Anim. Sci.* **49**, 319-323 (1999)

BC 1
Rat is emaciated
- Segmentation of vertebral column prominent if not visible.
- Little or no flesh cover over dorsal pelvis. Hipbones prominent if not visible.
- Segmentation of caudal vertebrae prominent.

BC 2
Rat is under-conditioned
- Segmentation of vertebral column prominent.
- Thin flesh cover over dorsal pelvis, little subcutaneous fat. Easily palpable.
- Thin flesh cover over caudal vertebrae, segmentation palpable with slight pressure.

BC 3
Rat is well-conditioned
- Segmentation of vertebral column easily palpable.
- Moderate subcutaneous fat store over pelvis. Easily palpable with slight pressure.
- Moderate fat store around tail base, caudal vertebrae may be palpable but not segmented.

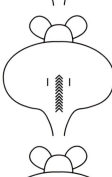

BC 4
Rat is over-conditioned
- Segmentation of vertebral column palpable with slight pressure.
- Thick subcutaneous fat store over dorsal pelvis. Palpable with firm pressure.
- Thick fat store over tail base, caudal vertebrae not palpable.

BC 5
Rat is obese
- Segmentation of vertebral column palpable with firm pressure; may be a continuous column.
- Thick subcutaneous fat store over dorsal pelvis. Not palpable with firm pressure.
- Thick fat store over tail base, caudal vertebrae not palpable.

BC=body condition

Figure 3B Rat body condition scoring. Body condition scoring is a way to assess the overall health and weight of an animal through sight and palpation.

Redrawn from Hickman, D. L. & Swan, M. Use of a body condition score technique to assess health status in a rat model of polycystic kidney disease. *J Am Assoc Lab Anim Sci* **49**, 155-159 (2010)

Figure 4 The nude mouse on the left has a chronic viral disease that has left it thin and small, compared to the mouse on the right.

Figure 5A The mouse on the left is obese, and weighs approximately three times as much as the mouse on the right. These mice have a mutation in a protein that controls weight and appetite.

Figure 5B The rat in the photo is overweight. This stock of rats carries a mutation that affects weight and appetite.

Figure 6A The nude mice pictured here are littermates. The one on the top is undersized when compared a normal animal.

Figure 6B The two C57BL/6 mice are littermates but one is undersized. The smaller weighs 5.9g, while the larger weighs 9.5g.

Figure 6C These two rats are the same age and the same stock (CD). The rat on the left has been allowed to eat whatever it wants, while the animal on the right has had its feed restricted. The yellowing of the fur of the back is normal in older rats.

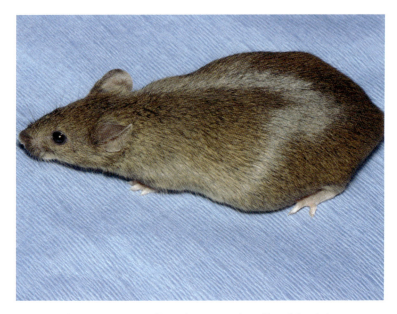

Figure 7A A pregnant mouse, illustrating a normal swelling of the abdomen. The mouse's coat is glossy and well-groomed, indicating that it is healthy.

Figure 7B This mouse has a grossly enlarged abdomen due to a blockage that resulted in urine retention. This animal also shows fur staining.

Figure 8A The mouse on the top shows a true hunching of the spine (arrow; kyphosis). This mouse carries a transgene that results in osteoporosis. It also has a rectal prolapse. The mouse on the bottom illustrates the normal "hunched" appearance seen after euthanasia of mice.

Figure 8B This animal is hunched and ruffled and is squinting; this is the typical appearance of a seriously ill mouse. (Photo courtesy of LAVA, UK; www.digires.co.uk.)

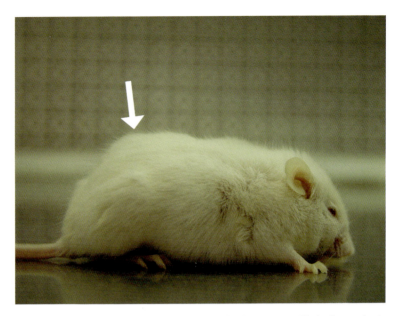

Figure 9 A hunched, ruffled mouse. Notice the body posture, with the feet tucked underneath, and the slitted eyes. The fur is ruffled and standing on end. The mouse is also pale. This mouse also has a large lump (arrow). This lump was a tumor.

Figure 10A The rabbit pictured here has a blue tint to the normally red eye and appears blue around the nose and whiskers. This animal had a heart defect.

Figure 10B The ears of the mouse on the left are pale when compared to the normal littermate.

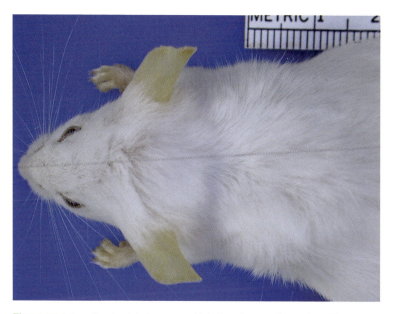

Figure 11 A jaundiced or icteric mouse. Note the strong yellow color of the ears and paws.

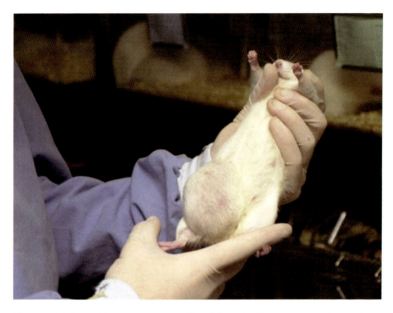

Figure 12A A rat with a mammary swelling. It is probably a mammary tumor since these tumors are very common in elderly rats.

Figure 12B A nude mouse with multiple swellings on its left side. These are tumors of the immune system which are common in mice, especially nude mice.

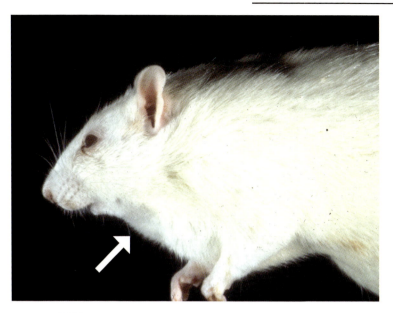

Figure 12C This rat has a swelling of the salivary glands under the neck (arrow), secondary to a viral infection (sialodacryoadenitis virus virus; SDAV).

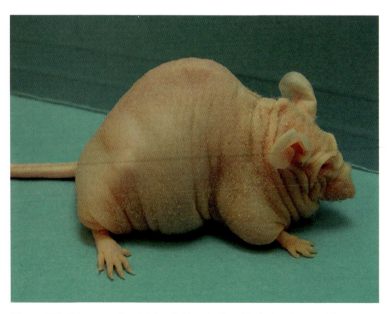

Figure 13A This mouse is retaining fluid under the skin (subcutaneously) throughout the body. This is most often seen secondary to severe kidney, heart, or liver problems.

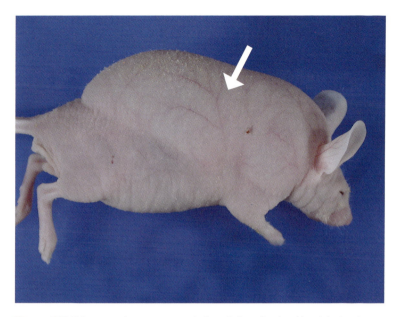

Figure 13B This mouse has an accumulation of air under the skin of the back. This is a rare problem that results from damage to the respiratory tract, either by gavage or by tumor formation.

Figure 14 This rat is showing an arching behavior associated with abdominal pain. This may be seen along with writhing or stretching. (Photo courtesy of Dr. P. Flecknell)

Figure 15A This rabbit has a broken back. Rabbits can easily fracture their backs if their strong hind legs are not held correctly.

Figure 15B This rabbit has a condition known as splayleg. This is usually a genetic disorder. (Photo courtesy of LAVA, UK; www.digires.co.uk.)

Figure 15C This mouse has paralyzed hind legs due to a tumor in the abdomen (not visible in this photo).

Figure 16A This nude mouse is dehydrated, as shown by the tented skin on its back.

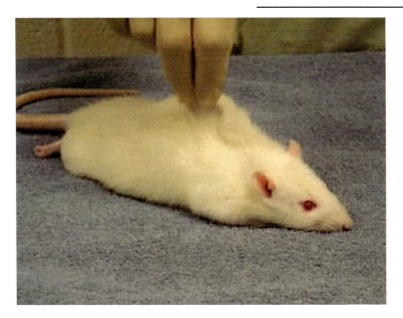

Figure 16B These two photos (16B and C) illustrate dehydration in a rat by showing how the skin will "tent" (stick together and stand up) when pinched (arrow in photo C).

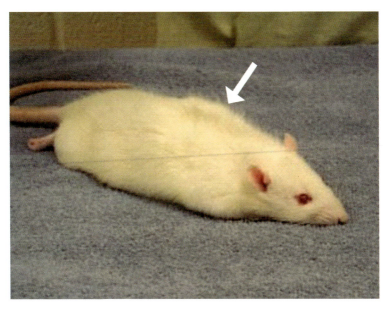

Figure 16C These two photos (16B and C) illustrate dehydration in a rat by showing how the skin will "tent" (stick together and stand up) when pinched (arrow).

Movement

Abnormal movements (Limping, paralysis, stereotyped behavior, seizures, spinning, flipping, twirling)

Description: An animal moves abnormally when compared to other animals of its strain, sex, and age.

Reasons: Animals may move abnormally because they are weak or have defects in their spine, legs or feet. Animals may move abnormally because they are having seizures (fits). Stereotypical behavior is a term referring to abnormal movements repeated over and over by some animals in captivity. In mice, these movements can include activities such as carrying the tail in the mouth for a long period or doing flips from the cage bars. When making an observation of abnormal movements in an animal, describe the type of movement, how often it is seen, and whether other animals are performing the same movements. (See *seizures, spinning/flipping/twirling, abnormal posture*)

Decreased activity (Lethargy, stillness, depression, unconsciousness, weakness)

Description: An animal shows decreased activity when compared to others of the same age, sex and genetic background. Normal animals move when disturbed and move away from discomfort.

Reasons: Decreased activity is a way for an ill or injured animal to conserve energy. Animals will also show decreased activity after administration of anesthesia.

Tremors (Vibrations, shakes)

Description: A tremor is a fine muscular movement. Animals may have whole-body tremors or tremors of body parts. Tremors may occur constantly or only when the animal is either moving or at rest.

Reasons: Animals may shake when cold. Tremors may also be due to certain neurological conditions, such as Parkinson's disease or birth defects (e.g., defects in the portion of the brain that controls movement).

Paralysis

Description: An animal cannot move. This condition should be distinguished from weakness or unconsciousness. Animals can be completely paralyzed on one half of their bodies, or more typically, from the mid-back down.

Reasons: Paralysis is usually due to damage to the spinal cord. This damage can occur because of experimental manipulations, illness, or injury. A cause of paralysis in rabbits is a broken back due to improper handling (failure to support the rabbit's strong hind legs). (See *abnormal movements*)

Straub tail

Description: A rat or a mouse has a stiffened, vertically raised tail. Normally, rats and mice carry their tails extended horizontally from their bodies when moving about the cage.

Reasons: This condition is caused by a muscle spasm at the base of the tail, seen when the nervous system is activated. This nervous system activation could be due to excitement, aggression, or the administration of certain drugs, such as morphine. It may be caused by administration of other compounds as well.

Spinning/flipping/twirling/circling

Description: An animal moves in circles, flipping over (either with the aid of the bars of the cage or falling over), or rolling over when it walks. An animal may also spin when picked up by the tail.

Reasons: Animals spin, flip, or twirl when they have problems with their balance. This condition may be a genetic problem, a problem with their brains, or an infection in their ears. They may also do flips using the bars of their cages or turn in circles as part of a pattern of disordered behavior sometimes seen in laboratory rodents (stereotypical behavior).

Staggering (Ataxia, wobbling, loss of balance)

Description: An animal staggers or walks in an uncoordinated manner. It leans from side to side as it walks and may even fall over. Sometimes, a staggering animal falls and cannot get back on its feet.

Reasons: Staggering is commonly seen in animals that are recovering from anesthesia or in the process of going under anesthesia. Staggering may also be a gait associated with weakness. Ear or balance disorders may also cause staggering, as may some spinal cord problems.

Seizures (Convulsions, fits)

Description: An animal having a seizure may show many different behaviors. Some animals look up at an awkward angle and twitch their heads. Others may fall over and lie on their sides, paddling or rhythmically contracting their legs (clonic). Another type of seizure is one in which animals may extend or contract their muscles and leave them in that position (tonic). In the tonic type, the animal may fall over on its side with its legs and head held rigidly away from or in toward the body.

Reasons: Seizures are caused by abnormal electrical activity in the brain. Many laboratory mice have seizures due to a genetic predisposition for seizures. For example, DBA/2 mice are prone to seizures caused by loud sounds.

Hair loss (Alopecia, shedding, barbering, overgrooming)
Figure 17

Description: Alopecia is the loss of hair, either all over the body or in a specific area. When describing hair loss, be sure to describe the size of the hairless area, the condition of the skin under the missing hair, the location of the hairless areas, and whether the hair loss is on both sides of the body.

Reasons: Alopecia may be the result of a particular event, such as barbering or hair pulling, or it may be caused by a generalized illness, such as cancer. Scratching due to parasites or some skin infections may also result in areas of hair loss. Alopecia may be caused by stress, such as rearing a litter of pups, or repeated pregnancies. It is commonly seen in older guinea pig breeder females. Some animals, such as rabbits, may also pluck out their fur to line nests.

Barbering (Hair chewing, overgrooming, whisker plucking)
Figure 18

Description: An animal displays bare patches of skin, usually on the back, shoulders or head. Where the fur is missing, the visible skin is often normal.

Reasons: This condition is most commonly seen in mice and is an abnormal behavior. One animal in the cage may appear completely normal (the barber), while all the others have patches of missing fur. Animals may also rub off the hair on their muzzles and heads on the bars of feeders. When this occurs, all animals in the cage tend to be affected. (See *hair loss*)

Shedding (Hair loss, alopecia)
Figure 19

Description: Shedding is the loss of hair through normal hair regrowth processes. The hair lost is regrown. It is abnormal if the hair does not grow back or the animal develops bare patches.

Reasons: Animals may shed if the room is too warm (this condition is often seen in rabbits). Animals may shed due to the season of the year, dietary problems, a hormonal imbalance, or the stress associated with giving birth and rearing a litter. (See *alopecia*)

Coat color changes
Figure 20

Description: All or part of the coat of an animal is not the normal color for that sex, age, or strain of animal. A careful description of the color change and how much of the coat is affected are important observations.

Reasons: Genetic contamination occurs when an animal of the wrong stock or strain enters a colony and contributes genetic material. If genetic contamination causes changes in coat color, the entire coat should be affected and more than one animal in a litter may have an abnormal coat color. The consequences of such a mix-up may not be seen for several generations. Other indicators of genetic contamination include abnormally large litter sizes or larger than usual pups. Coat color changes may also occur due to a spontaneous genetic mutation in one animal in a litter. Finally, after an injury, the coat of an animal may grow back in a different color than it was before. Animals also gain white or gray hairs in their coats as they age. This condition will be most visible in animals with darker coats.

Belly spot (White spotting)
Figure 21

Description: A patch of white hair appears on the belly of a pigmented animal. It can be a small spot of only a few white hairs to an extensive area of the belly, spilling over onto the sides of the animal.

Reasons: Belly spots are caused by the mutation of a gene that controls the movement of the cells that contain pigment. When an abnormality in this gene occurs, these cells do not move to the midline of the rodent, leaving a white patch. (See *coat color changes*)

Curly or wavy coat
Figure 22

Description: An animal has a curly or wavy coat of hair.

Reasons: This condition occurs when there is a mutation in the animal's genetic material. It results in abnormal hair and whisker growth. This condition may occur in rats, rabbits, and mice. It has not been identified in gerbils, guinea pigs, or hamsters.

Head tufts
Figure 23

Description: A small tuft of hair appears on the top of an animal's head. This fur often faces backward from the rest of the fur or straight up and down.

Reasons: This condition does not seem to be an inherited trait. It is more commonly seen in C57BL/6 mice. The reason behind the occurrence of head tufts is unknown.

Fur staining (Stained hair coat, dirty, urine-stained, greasy)
Figures 6C, 7B, 24, 61, 69

Description: An animal has a hair coat that is stained or dirty in appearance. Fur staining will be more obvious on a white or light-colored animal. When describing stained fur, note the location of the stained area and the color of the stain.

Reasons: An animal's hair may be stained with feed or bedding; with discharges from the eyes, nose, ears or mouth; with fecal matter; or with urine. Some animals may change color slightly as they age—white animals may become more yellow. Animals may become stained from the products of an illness, such as diarrhea. Animals may also appear stained or greasy when they are not grooming themselves properly. This lack of grooming may be due to illness or pain.
(See *ruffled, nasal discharge, ocular discharge, ear discharge, drooling*)

Figure 17 This guinea pig has lost all her hair due to hormonal changes associated with repeated pregnancy. She also has some small scratches on her sides due to scuffles with cagemates.

Figure 18A These photos (Figures 18A-18E) illustrate various types of barbering. This can be a significant problem for female C57BL/6 mice. A barbered pup is on the left and a normal one on the right. The dam has plucked all the fur from one pup.

Figure 18B These photos (Figures 18A-18E) illustrate various types of barbering. This can be a significant problem for female C57BL/6 mice. This mouse has had its whiskers barbered.

Figure 18C These photos (Figures 18A-18E) illustrate various types of barbering. This can be a significant problem for female C57BL/6 mice. This group of animals has been barbered to various degrees. It is not always possible to identify the barber in a group, as animals may self-barber as well as barber each other.

Figure 18D These photos (Figures 18A-18E) illustrate various types of barbering. This can be a significant problem for female C57BL/6 mice. Another barbered animal. In this case the whiskers are missing, and the fur is plucked to bare skin on the flank. (Photo courtesy of Dr. J. Garner)

Figure 18E These photos (Figures 18A-18E) illustrate various types of barbering. This female rat has pulled all the fur from her belly and legs, either due to barbering behavior or to line a nest for her pups.

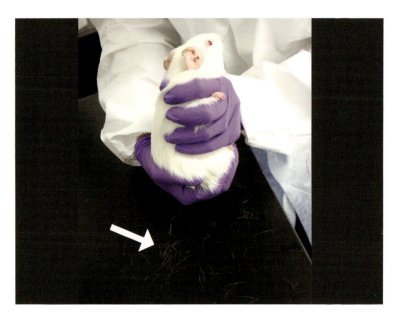

Figure 19 This guinea pig is illustrating the normal shedding which occurs when animals are handled. Note the white fur on the black countertop.

Figure 20A In this litter of albino rats, one has spontaneously "repaired" the albino mutation and reverted to wild-type.

Figures 20B, C, D The male mouse in this cage is a DBA/2. He has a spontaneous mutation that results in a normally-colored back but a pale belly. Four of his six offspring shown also have the mutation.

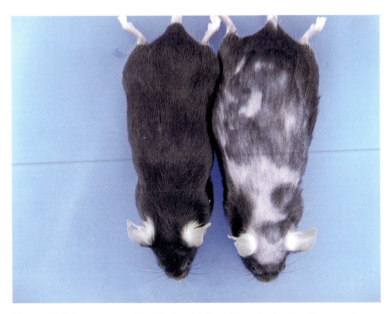

Figure 20E The mouse on the right is a barbered female showing the normal appearance of C57BL/6 mouse skin as the fur regrows. Although the skin appears to have black patches, it is actually the new hairs underneath the white skin that give it the appearance of a pigment change.

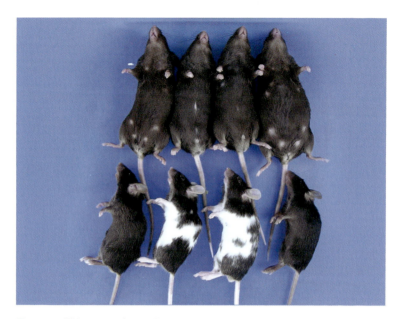

Figure 21 This group of mice illustrates a relatively common mutation that occurs in mice. The two adults in the center of the top row both have small patches of white on their bellies. This mutation passes on to offspring, resulting in the white spotting seen in the pups in the bottom row.

Figure 22 This rat carries the Rex mutation, which results in animals with wavy or curly hair. (Photo courtesy of Dr. T. Kuramoto; Kuramoto, T., Hirano, R., Kuwamura, M. & Serikawa, T. Identification of the rat Rex mutation as a 7-bp deletion at splicing acceptor site of the Krt71 gene. *J Vet Med Sci* **72**, 909-912, (2010).)

Figure 23 This C57BL/6 mouse has a tuft of hair growing from its head.

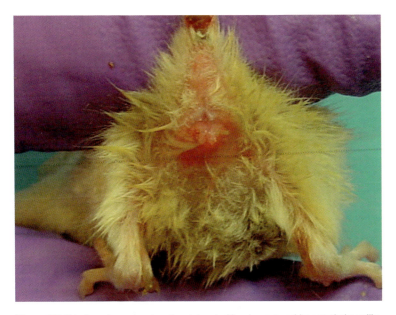

Figure 24 This female mouse has fur stained with urine around her genital papilla.

Thick skin
Figure 25

Description: A mouse with no hair has skin that appears to be thicker than normal. The thickness of the skin is often assessed on the back, between the shoulder blades. This condition can occur in hairless or nude mice.

Reasons: Mice without hair still have the cells that produce hair, but the hairs produced are abnormal. These cells go through the normal hair cycle of hair growth and loss. When hairs are being formed, the skin will appear thicker. This condition is normal for the nude mouse.

Thin skin
Figure 25

Description: Thin skin is the normal appearance of mice such as nude or hairless. When looking at a nude or hairless mouse, you should be able to see some of its internal organs as shadows beneath the skin of its belly.

Reasons: This condition is normal for the nude mouse.

Scaly skin (Hyperkeratosis)
Figure 26

Description: This condition mainly affects nude mice, where parts or all of the skin are covered with numerous, small, white or yellowish flakes. The mouse may even look like it was dipped in cornmeal or sand.

Reasons: This condition may be caused by a serious bacterial infection. It might also be caused by a lack of grooming if the mouse is ill.

Fight wounds (Bite wounds)
Figures 17, 27, 71

Description: Wounds appear on an animal that are caused by the teeth and claws of another animal. These wounds most often appear on the tail, the hind feet, and the rump of the animal. Fight wounds may also appear on the belly. One animal in the cage is usually fine, while the others are wounded. In animals with no hair, the wounds are usually more visible.

Reasons: Fighting is more common in male animals and is usually related to establishing dominance in the group. The dominant animal in the cage usually has no fight wounds.

Abrasion (Scrape)
Figure 28

Description: An abrasion is an area on the skin of an animal where the outer layer has been rubbed off. It can be red or pale and may weep clear fluid.

Reasons: Abrasions occur when animals injure themselves or each other by rubbing against something.

Discolored skin (Bruising, red spots, white spots, pigment changes)
Figure 29

Description: A change occurs in the color of the skin of an animal. The color change can be from normal to red, purple, yellow, green, white or any shade in between.

Reasons: An animal's skin may be discolored for many reasons, such as injury, scarring, allergy or disease. Animals may have red, raised (rashy) areas of their skin for no reason that can be determined. Sometimes these areas are seen after handling. Animals may have bruises after blood collection. Some diseases of pigmentation result in changes in the color of an animal's skin.

Ulceration
Figure 30

Description: An ulceration is an open sore. The first layer of skin is gone, and the underlying layer is visible. Ulcers may be so deep as to show muscle or bone, but this is rare. They often weep clear fluid. Ulcerations may have a scab.

Reasons: Ulcers may appear if animals injure themselves through scratching or overgrooming. Burn wounds often appear as ulcerations. Some bacterial infections may appear as ulcerations. Ulcerations may occur when a tumor outgrows its blood supply and dies (you may see an ulcerated lump, for example). (See *ulcerative dermatitis*)

Ulcerative dermatitis (B6 dermatitis)
Figure 31

Description: Animals with ulcerative dermatitis have areas of raw or missing skin on parts of their bodies. The red, wet appearance of the ulcer is because the external protective layer of skin is gone, revealing the tissue underneath. Over time, the ulcer may dry and scab or even scar, damaging the animal's skin. Ulcerative dermatitis may be seen on animals' backs or flanks, or on their heads, necks and ears. It is less commonly seen on the belly, feet or tail.

Reasons: Dermatitis is a general term that means "inflammation of the skin." Ulcerative describes how it looks; it is raw and open (ulcerated), not dry or scaly. Dermatitis may be caused by contact with chemicals that damage or burn the skin, such as certain disinfectants. Dermatitis may also be caused by fighting or injury due to a rough cage surface. If these wounds become infected, they may lead to ulcers. Ulcerative dermatitis can be caused by the scratching associated with fur mite infestation. In some strains of mice, such as C57BL/6, ulcerative dermatitis is occasionally found with no underlying cause or secondary to barbering. (See *ulceration, barbering*)

Fissure

Description: A fissure is a deep crack or crevice in the skin.

Reasons: Fissures may occur when the skin is too dry.

Laceration (Cut)
Figures 17, 51

Description: A laceration is a cut or tear of the skin. A laceration involves the deeper layers of the skin, unlike a scratch. The skin around a laceration is often normal, or slightly reddened, but is not dry like it would be around a fissure.

Reasons: Lacerations occur when animals encounter sharp objects, such as the teeth or claws of other animals, a sharp cage edge, or a needle or scalpel. (See *fight wound*s)

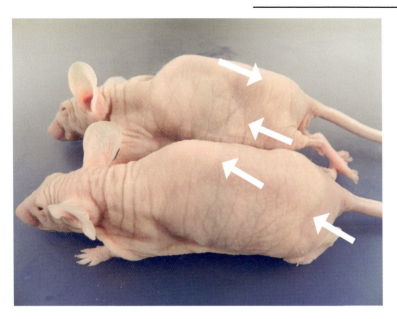

Figure 25 These nude mice are illustrating the apparent thickening of skin seen with the failed hair cycle associated with the nude mutation. The active hair follicles appear as lines (arrows) moving down the animal from nose to tail.

Figure 26A These nude mice (26A and B) have scaly skin. This condition is caused by infection with *Corynebacterium bovis*.

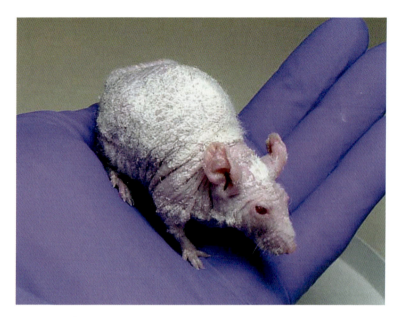

Figure 26B These nude mice (26A and B) have scaly skin. This condition is caused by infection with *Corynebacterium bovis*.

Figure 27A Mild healing bite wounds on the back and rump of a male nude mouse.

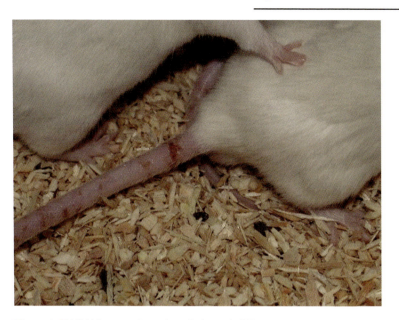

Figure 27B Mild bite wounds on the tail of a male OF1 mouse.

Figure 27C Severe bite wounds on the rump of a male albino mouse.

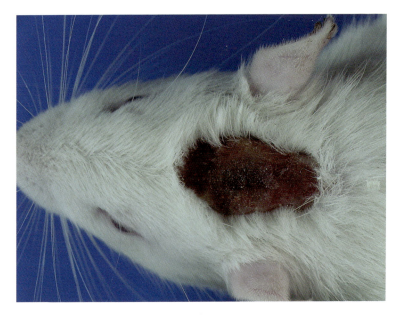

Figure 28 An abrasion on the head of a CD rat.

Figure 29 Discolored skin on the abdomen of a male nude mouse. This discolored skin is due to bruising.

Figure 30 This nude mouse has an ulcerated xenograft (foreign tissue implanted into the mouse; probably a tumor line).

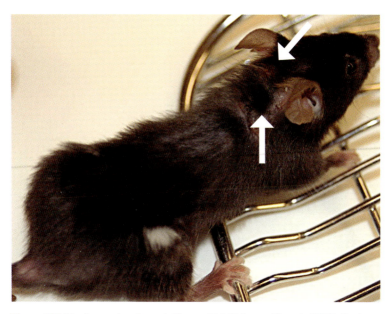

Figure 31A The three mice shown in Figures 31A-31C are all female C57BL/6 mice. Mild ulcerative dermatitis on the shoulders and between the ears.
(Photo courtesy of Dr. J. Garner)

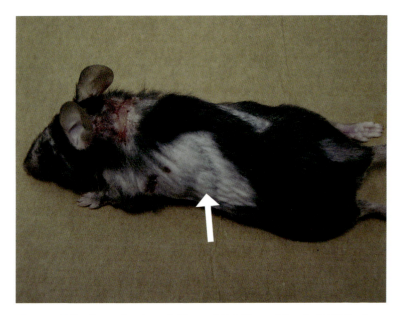

Figure 31B The three mice shown in Figures 31A-31C are all female C57BL/6 mice. Moderate ulcerative dermatitis between the shoulder blades and behind the ears. This animal also illustrates barbering and hair regrowth (arrow).

Figure 31C The three mice shown in Figures 31A-31C are all female C57BL/6 mice. This animal has a severe ulcerative dermatitis. These lesions are often complicated by secondary bacterial infections of the ulcerated skin.

Head tilt
Figure 32

Description: An animal's head tilts to one side.

Reasons: The most common reason for a head tilt in an animal is an ear infection. Less common reasons include damage to the brain or inherited changes in the structures in the ear. These animals may only move to one side in the cage, or they may spin in circles when you pick them up by the tail. (See *abnormal movements*)

Domed head (Hydrocephalus, water on the brain)
Figure 33

Description: An animal has a domed head, usually accompanied by squinting eyes, small size, and abnormal movements. It is usually seen in animals at weaning, or not long after.

Reasons: The brain and spine are surrounded by fluid. This fluid is produced by the brain and circulates freely through passages in the brain and spinal cord. When fluid passages in the brain are blocked, the fluid builds up in the brain, causing swelling of the skull and compression of the brain tissue. In both rabbits and rodents, hydrocephalus has a genetic component. In rabbits, hydrocephalus may be also be caused by a lack of certain vitamins in the diet.

Overgrown teeth (Malocclusion, slobbers)
Figure 34

Description: An animal's teeth are too long. They are often also not aligned properly.

Reasons: The teeth of rodents and rabbits grow constantly. The teeth wear on each other and on the hard chow the animals are fed. If the teeth or jaw of the animal is damaged or misaligned, the teeth continue to grow and do not wear properly. In mice, rats and rabbits, the malocclusion usually occurs in the front teeth (incisors). Powdered or soft diets may predispose animals to overgrown teeth. In guinea pigs, it is more commonly seen in the back teeth (molars). In rabbits, this condition is often a problem inherited from the parents. In rodents, it is probably partially due to heredity. (See *undersized, drooling*)

Swollen face
Figure 35

Description: An animal has a snout that is larger than expected and puffy.

Reason: Animals may show a swollen face when they have misaligned teeth or a broken jaw. In animals housed on wire or punched-metal bottom cages, animals may get their teeth caught in the cage bottoms. This situation results in damage to the upper or lower jaw. (See *overgrown teeth*)

Drooling (Excessive salivation)
Figure 36

Description: An animal releases saliva in amounts that wet and stain the area around its mouth.

Reasons: Animals may seem to salivate excessively when they cannot swallow their saliva. Animals may also drool when their mouths or throats are painful. Drooling may indicate malocclusion. In rats and mice, drooling is sometimes seen when animals are overheated. (See *overgrown teeth, fur staining*)

Nasal discharge (Runny nose, snuffles)
Figure 37

Description: An animal releases fluids from its nose. Nasal discharge can be red, clear, yellow, white, or green. The discharge can also be described as thin or thick.

Reasons: Animals may have nosebleeds (rare) or nasal discharges associated with nasal, sinus, eye, or lung infections. They may also have discharges if something in their environment is irritating the lining of the nose or they have a foreign body in their nose. Rats, mice, gerbils, and hamsters also have a gland behind their eye (the Harderian gland) that normally produces a substance called porphyrin. When exposed to light, this substance darkens and looks like blood. This gland always produces this substance, but when animals are sick or stressed and stop grooming, the secretions will build up and secretions can crust at the rims of the eyes and the edge of the nose. Porphyrin will glow under a black light and blood will not. In rats, a red nasal discharge may indicate a serious viral infection, although it is usually due to stress or dehydration.

Sore nose

Description: A gerbil has a red, sore nose. Crusting and oozing of the skin around the nose may occur.

Reasons: Gerbils have glands behind their eyes (the Harderian glands) that produce a substance (porphyrin) that can irritate their skin if not regularly removed by grooming. Gerbils get sore nose when they do not groom regularly, which may be due to illness or other problems. (See *nasal discharge*)

Sneezing

Description: An animal sneezes.

Reasons: Animals may sneeze due to infection or other illness or due to irritating substances in their environment. (See *abnormal breathing*)

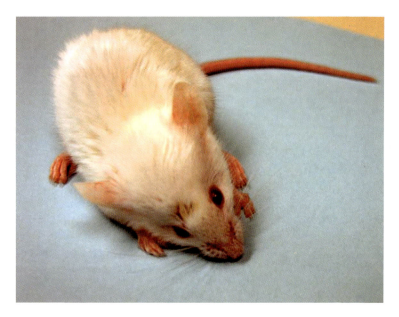

Figure 32A A mouse with a head tilt.

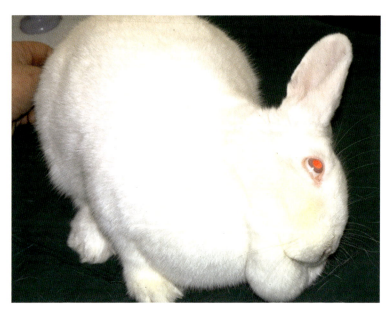

Figure 32B A New Zealand White rabbit with a head tilt.

(Photo courtesy of LAVA, UK; www.digires.co.uk.)

Figure 33A The rat on the top has a domed head and is much smaller than the littermate on the bottom. The runted rat weighed 83g while the normal littermate weighed 131g.

Figure 33B The mouse on the right has a domed head when compared to the littermate mouse on the left.

Figure 34A The normal occlusion ("bite") of the mouse. The bottom teeth appear longer than the top and the greatest wear occurs as the top and bottom teeth wear against each other.

Figure 34B A mouse with overgrown bottom teeth.

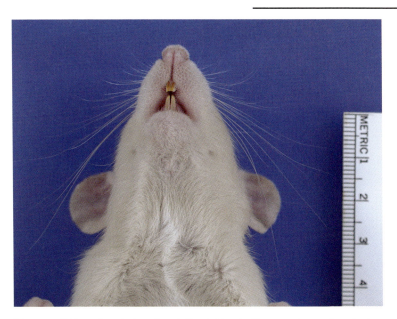

Figure 34C Normal occlusion of a rat. The bottom teeth appear longer than the top and the greatest wear occurs as the top and bottom teeth wear against each other.

Figure 34D A malocclusion in a rat. This is a relatively severe case.

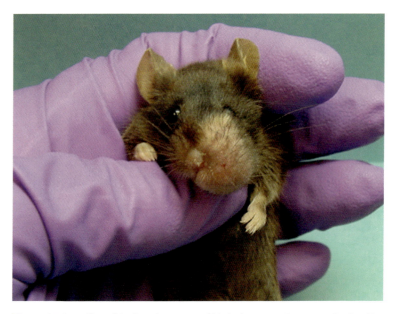

Figure 35 A swelling of the face in a mouse. This is due to an abscess under the skin.

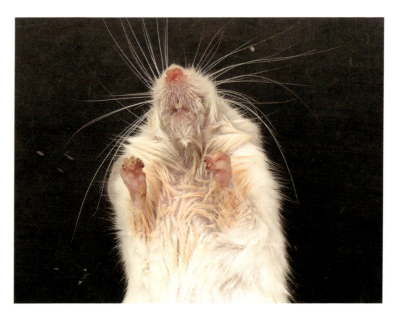

Figure 36 A mouse that overheated. Since mice and rats cannot sweat, it wetted its fur with saliva in an attempt to cool itself.

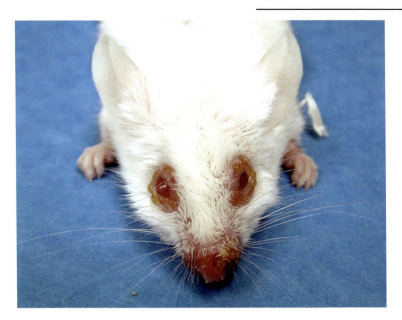

Figure 37A Ocular and nasal discharge in a mouse.

Figure 37B Nasal discharge in a rabbit. (Photo courtesy of LAVA, UK; www.digires.co.uk.)

Dilated pupils (Mydriasis)
Figure 38

Description: An animal's pupils are more dilated than expected for the light conditions (in dim light, pupil dilation is normal). Also, pupils do not constrict when exposed to bright light.

Reasons: Animals can have dilated pupils due to the administration of certain drugs or compounds. These compounds may have been administered systemically or directly into the eye. Animals may also have dilated pupils with brain or nerve damage. Diseases of the eye may also result in dilated pupils. It may be difficult to see changes in the pupils of small animals, especially those with dark pupils.

Constricted pupils (Miosis)
Figure 39

Description: An animal's pupils are smaller than expected for the light conditions (it is normal for pupils to be constricted in bright light). Also, pupils do not dilate when the animal is put in dim light.

Reasons: Animals can have constricted pupils due to the administration of certain drugs or compounds. These compounds may have been administered systemically or directly into the eye. Animals may also have constricted pupils with brain or nerve damage. Diseases of the eye may also result in constricted pupils. It may be difficult to see changes in the pupils of small animals, especially those with dark pupils.

Different-sized pupils (Anisocoria)
Figures 38, 39

Description: An animal has pupils of different sizes (e.g., one dilated and one normal).

Reasons: Different-sized pupils may be an indication of a significant head injury or simply an injury to one eye. It may also be seen with certain nervous system injuries.

Spots on or in the eye (Cloudy eyes, corneal opacities, lenticular opacities, cataracts)
Figure 40

Description: An animal has white spots on its eye (corneal opacities) or in its eye (lenticular opacities).

Reasons: Cataracts (or lenticular opacities) are seen when the normally clear lens within the eye is opaque. Cataracts may be present from birth, they may be due to trauma, or they may be due to another underlying condition, such as age or diabetes. Corneal opacities are seen when the cornea, the clear covering over the pupil of the eye, becomes opaque. This is usually due to damage to the cornea, but it may be a common strain-related finding in mice and rats. If the cornea was injured, then it often heals with whitish scars.

No eyes (Anophthalmia)
Figure 41

Description: An animal is missing either one or both of its eyes.

Reasons: Anophthalmia is often a birth defect. A small percentage of newborn animals of some mouse strains, especially C57BL/6, show this abnormality. Occasionally, it occurs after an infection or damage to the eye. It may be difficult to tell anophthalmia from micropthalmia.

Small eyes (Microphthalmia)
Figure 42

Description: An animal has one or both of the eyes smaller than the eyes of others of its age and size.

Reasons: Microphthalmia is usually a birth defect. It is relatively common in the C57BL/6 strain. Microphthalmia may also be seen after an eye infection that damages the eye. It may be hard to tell anophthalmia from micropthalmia without specialized examination of the eye.

Large eyes ("Bug eyes," glaucoma, exophthalmia, megaloglobus, hypopion, buphthalmia)
Figures 43, 44

Description: The eyeball of the animal is larger than normal.

Reasons: Some animals, like rats and rabbits, tend to have eyes that stick out slightly. This condition is normal for these animals and is due to a shallow eye socket. This protrusion, or pushing out, of the eyeballs can be made worse by handling the animal. Exophthalmia (protrusion of the eye from the socket) is also produced when there is a problem behind the eye that pushes the eye forward, out of the socket. Exophthalmia may also be seen when the eye is infected due to trauma or other causes. It is often seen in nude mice with bacterial infection of the glands behind the eye. Megaloglobus is an overall enlargement of the eyeball, usually caused by a defect in the circulation of the fluid in the eye (glaucoma). This condition is most commonly seen in rabbits, where it is called buphthalmia. Buphthalmia may be associated with nutritional problems.

Sunken eyes (Enophthalmos)
Figure 45

Description: The eye of an animal is pulled back into its socket and often appears small and dull. The third eyelid may be visible, if the animal has one.

Reasons: Sunken eyes may be due to dehydration, general illness, or damage to a part of the nervous system. Sunken eyes may be difficult to detect in the smaller species of rodents.

Squinting
Figures 9, 10B, 46

Description: An animal's eyes are held closed or almost closed. The eyes may appear to be slits.

Reasons: Animals may squint if the light is too bright or if their eyes are painful. Animals may also squint if they have pain or are ill. (See *swollen or red eyelids, small eyes, no eyes*)

Swollen or red eyelids (Squinting, conjunctivitis, blepharitis)
Figures 46, 47

Description: The eyelids of an animal are swollen, reducing the part of the eye visible to a slit, or nothing. In addition, the eyelids are often red. The eyelids may appear swollen when they are held closed or only opened slightly by the animal.

Reasons: Animals may have swollen eyelids due to trauma to the eye or eyelid, infection, or compound administration. In hairless animals, it often indicates that bits of bedding or small hairs have built up under the eyelid. Animals may also hold their eyes shut if they are in pain or are more sensitive to light due to compound administration or inflammation. (See *ocular discharge*)

Drooping eyelid (ptosis)

Description: An animal has a drooping upper eyelid. It will make the affected eye appear smaller or slitted.

Reasons: Animals usually show drooping eyelids when they have damage to a particular part of their nervous system. This condition may be difficult to detect in small rodents.

Ocular discharge (Runny eyes, pus in the eye)
Figures 46, 47, 48

Description: Tears, pus or other substances issue from the eye of the animal. Production of some tears is normal, but tearing or discharge that stain the face or paws is not normal.

Reasons: Ocular discharge may be present for a variety of reasons. Infectious disease, either of the whole animal or just of the eye, may result in an ocular discharge. Animals may have ocular discharge if there is something irritating in the eye. Ocular discharge may also be secondary to trauma or administration of an irritating compound. In rabbits, ocular discharge can be due to a blocked tear duct. (See *swollen or red eyelids*)

Red tears (Chromodacyrorrhea)
Figures 49, 69A

Description: An animal has a reddish discharge or staining around its eyes. It may appear as a faint pinkish stain or as reddish crusting. It may closely resemble blood. This condition is rarely seen in rodents other than rats.

Reasons: Rodents have special tear glands, called Harderian glands, located behind their eyes. These glands drain into both their eyelids and their noses. The Harderian gland produces porphyrin, a reddish pigment that looks like blood. Animals normally produce this secretion, and when they are healthy, they groom it away. When stressed, animals do not groom as well or as often. The stress may be due to many reasons, including illness. (See *nasal discharge*)

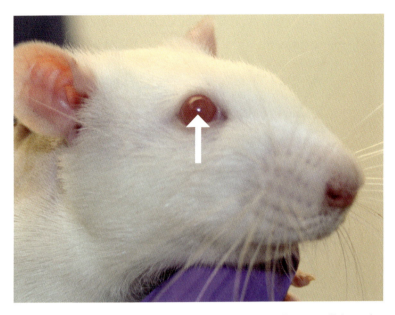

Figure 38 The arrow points to a normal pupil in a rat. Pupil diameter will depend on light levels. In dim light, pupils will be larger (dilated).

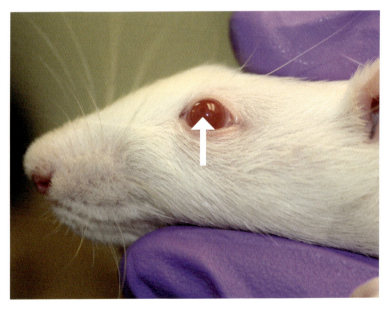

Figure 39 A constricted pupil in a rat (arrow). This is the same rat as in Figure 38, illustrating that pupils can be different sizes in the same animal. This animal suffered damage to a nerve, resulting in these clinical signs.

Figure 40A A spot on the surface of the eye (the cornea) of a mouse.

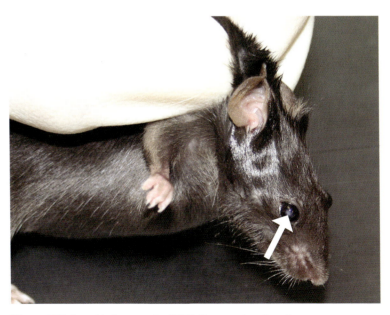

Figure 40B A spot in the eye of a C57BL/6 mouse (a cataract).

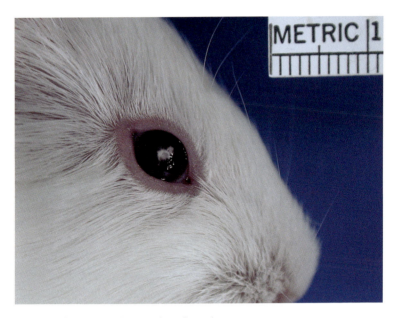

Figure 40C A spot on the eye of a guinea pig.

Figure 41 The mouse on the left is normal. The mouse on the right was born without eyes (anophthalmia). This is a relatively common condition in C57BL/6 mice. The eyelids are fused shut and cannot be opened.

Figure 42A The mouse on the left is normal. The mouse on the right has one eye that is smaller than normal (arrow; microophthalmia). This is a relatively common condition in C57BL/6 mice.

Figure 42B These euthanized mice are the same as shown in the figure above. The eyelids of the mouse with microophthalmia are held open with forceps to illustrate the small eye present in the socket.

Figure 43 This rat's left eye is larger than its right. This could be due to an infection or a problem with the pressure in the eye.

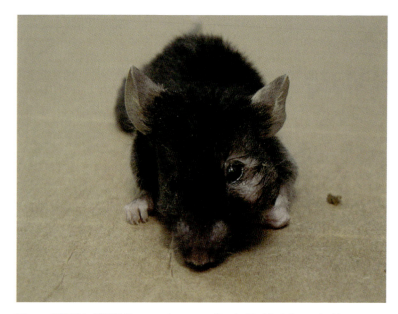

Figure 44A This C57BL/6 mouse has a swelling behind its left eye. In this case, this was due to an infection of the structures behind the eye.

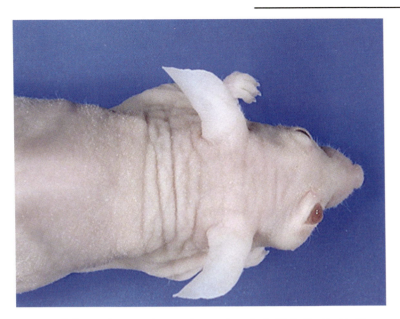

Figure 44B This nude mouse has an abscess (a walled-off infection) behind its right eye.

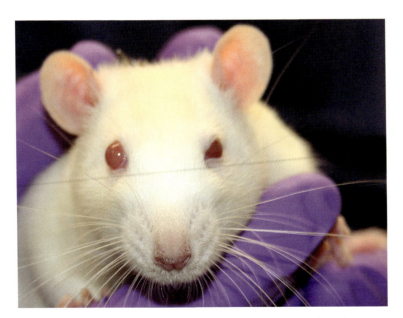

Figure 45 This rat is the same as in Figures 38 and 39. The rat's left eye is retracted into the socket, and the right eye is normal. It is normal for the eyes of rats and mice to protrude slightly from the socket.

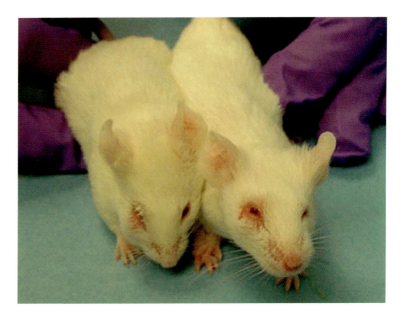

Figure 46 These mice have bacterial infections of the eyes. They are squinting, and have reddened eyelids. In addition, the mouse on the left has a swelling beneath its right eye.

Figure 47 This nude mouse has swollen eyelids, and also has an ocular discharge.

Figure 48 This nude mouse has an ocular discharge, but the eyelids are not swollen. Nude mice have no eyelashes, so they are prone to getting foreign material, such as bedding, in their eyes.

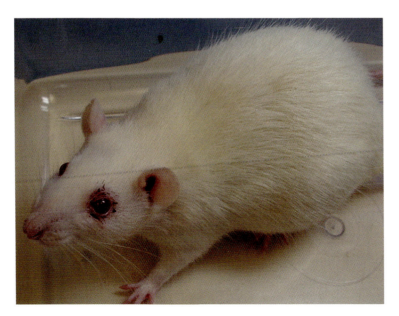

Figure 49 This albino rat is showing chromodacryorrhea or red tears. This reddish pigment is produced by the Harderian gland behind the eye. It is a general sign of stress or illness in rats.

Ears

Small ears
Figure 50

Description: An animal's ears are smaller than normal for the age and size of the animal.

Reasons: This condition may be caused by a mutation, but it may also be a result of ear chewing in mice or rats.

Ear injury (Torn ears, cauliflower ear, chewed ears, swollen ears, auricular chondropathy)
Figure 51

Description: An animal has ears that are ragged at the edges due to chewing by other animals in the cage or damage caused by torn ear tags or punches. In rats, the ears may appear thickened and swollen. In mice, the edges of the ears may become reddened, ulcerated, and may slough off. In rabbits, the ear veins and arteries are used to administer compounds and take blood. If the circulation is damaged during procedure, the rabbit's ears may be hot to the touch, then turn red and black and, in extreme cases, parts of the ear may slough off.

Reasons: Ear chewing in guinea pigs is a dominance behavior, one that is learned from other animals. Chewed ears may also result from fighting in rats and mice. Frequent fighting is a characteristic of some strains of inbred mice. In rabbits, ear trauma is often due to damage to the ear arteries or veins through administration of substances. Some rats may develop a condition due to their body reacting to the cartilage in their ears. These ears become thickened and lumpy and look like "cauliflower ear" in humans. Reactions to ear tags can result in thickened lumpy cartilage, holes in the ears, and infections. This condition may be associated with ulceration of the neck and shoulders and can affect one or both ears. (See *ulcerative dermatitis*)

Ear discharge (Crusty ears, ears with pus)

Description: An animal's ears have a discharge or crusts inside.

Reasons: Crusty ears, especially in rabbits, can indicate ear mites, but may be due to other causes. In rodents, crusty ears may be seen secondary to an ear infection associated with an ear tag or with trauma. In rare cases, infections inside the ear can lead to a discharge that spills out onto the fur. Rats also have a gland right beneath the ear (Zymbal's gland) that may become cancerous. This condition might look like an ear discharge.

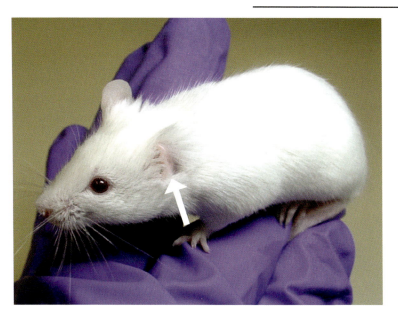

Figure 50 This mouse is missing an ear (arrow). It may be genetic, or due to a previous injury.

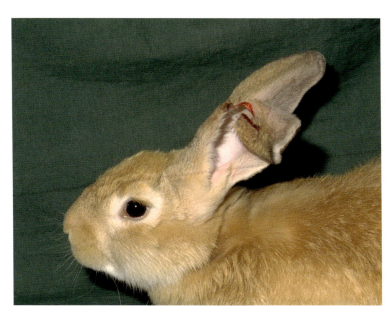

Figure 51A This rabbit has an ear injury. Ear injuries in rabbits may be due to fighting, ear tags, or improper administration of substances into the marginal veins or central artery. (Photo courtesy of LAVA, UK; www.digires.co.uk.)

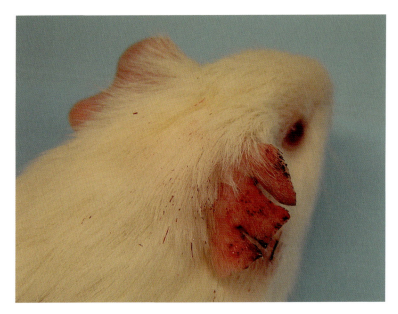

Figure 51B This guinea pig shows ear injuries associated with fighting.

Normal feces
Figure 52

Description: Fecal matter is a combination of undigested or indigestible material the animal has consumed and bacteria from the animal's intestines. The shape and character of the feces are specific to each species. For rabbits, this means two different types of feces. One type is dry and hard and produced during the day. The other type is moister, clumps together, and is produced at night. Rabbits must eat these night feces, which provide them with water, fiber, and essential vitamins. For that reason, they are rarely found in the cage pan. Rodents differ widely in their normal fecal output and appearance. Most rodents will eat some of their feces, and young rodents eat adult feces to build up their digestive bacteria.

Soft feces
Figure 53

Description: In many rodents, the loosest stool seen is a larger, softer, sticky stool, rather than liquid diarrhea.

Reasons: See *diarrhea*.

Loose feces

Description: Feces are loose, but not liquid. They are less formed than normal or soft feces, but more formed than diarrhea. Cow feces (cow pats) are a good example of loose feces.

Reasons: See *diarrhea*.

Diarrhea (Wet tail)
Figure 54

Description: An animal has copious liquid stool. This condition is less common in rodents than soft stool. Rabbits may have liquid diarrhea that sticks to their fur.

Reasons: Diarrhea may be caused by infectious disease. It may also be due to a change in diet or a genetic problem with the intestines.

No feces (Constipation, low fecal output)

Description: An animal may retain fecal matter, which is called constipation. It may also produce small amounts of hard, dry feces.

Reasons: Constipation may be due to a diet change or problem. It may also occur when animals do not consume enough water or feed. When fecal output is decreased, always check the amount of food and water being consumed. A lack of feces may also be seen in serious conditions, such as intestinal blockages or nervous system damage that affects the nerves to the intestines. (See *thin*)

Discolored feces (Bloody feces, yellow feces, gray feces, greasy feces)

Description: An animal's fecal material is an unexpected color. Feces are usually brown.

Reasons: Fecal material can change color when the digestion is disturbed. If there is blood being shed in the digestive tract, the feces can appear black (digested blood; higher in the intestines or stomach) or bloody (undigested blood; lower in the intestines). Animals that are jaundiced may have yellow stool, and animals with problems absorbing food may have feces that appear greasy or gray.

Rectal prolapse
Figure 55

Description: An animal has pink or red tissue sticking out of the anus.

Reasons: A prolapsed rectum occurs when an animal pushes a small portion of the last part of its large intestine outside of its body. Prolapse of the rectum is more common in mice than in any other species, but also occurs in guinea pigs. It is usually due to straining from giving birth, an intestinal infection, or constipation.

Normal urine
Figure 56

Description: Urine is a product of the kidneys, which are organs that help the body to filter waste from the blood. It is held in the bladder and passed from the body through the urethra. Normal urine varies in amount, color and clarity from species to species. Some species, like mice and rats, produce clear urine. Other species, like rabbits and hamsters, produce opaque, mucoid urine with many minerals in it. Rabbit urine may be yellow, yellowish-brown, yellowish-white, or even yellowish-red and still be normal. Some desert rodent species, like gerbils and hamsters, produce less urine when compared to other rodent species. Check the bedding first for normal urine color. To see if animals are producing abnormal urine, it may be necessary to put them in an empty cage or on white paper.

Bloody urine (Hematuria)

Description: An animal passes urine with blood in it. The blood can give the urine a pinkish tinge, or it can be present in clots or strings.

Reasons: Blood may be found in the urine for many reasons. It may be due to damage to the kidneys, bladder, urethra, or genitals. In rats, it usually indicates either infection of the bladder (cystitis) and kidneys or a birth defect of the kidneys called hydronephrosis. It may also be due to cancer in any of those organs or due to a problem that makes the animal's blood clot more slowly than normal.

Straining to urinate (Dysuria, stranguria)

Description: An animal has difficulty emptying the bladder of urine. The condition is usually accompanied by frequent urination and by pain on urination.

Reasons: Animals may strain to urinate if their urethra is blocked or if they have bladder infections. The urethra can be blocked due to trauma (to the penis, for example), or the animal may have bladder stones that are blocking the exit. (See *too little urine, dribbling urine, penis stuck in the prepuce*)

Dribbling urine (Urinary incontinence)
Figures 7B, 15C, 24, 59

Description: An animal leaks urine from the bladder.

Reasons: Urinary incontinence may be seen with damage to the spinal cord that leaves the animal unable to empty its bladder. Urine leakage may also be seen if the bladder contains stones or with a urinary tract infection. Urine dribbling may also be seen if the bladder is too full and cannot be emptied well. (See *fur staining, paralysis, penis stuck in the prepuce*)

Pale urine

Description: An animal passes urine that is paler than normal for its sex, age and strain. This condition means that the body is excreting a great deal of water, which is diluting the normal waste products.

Reasons: Pale urine may be seen if an animal is drinking more than it needs. This excessive drinking may be due to diabetes, problems with its brain or hormone levels or kidney problems. It may also be due to giving the animal too much fluid during a procedure. Pale urine may be difficult to see in smaller animals. (See *too much urine*)

Too much urine (Polyuria)

Description: An animal has an increased volume of urine for its age, sex and strain.

Reasons: The most common reason for an increased output of urine is diabetes. The increased sugar level in the blood in diabetes makes animals thirsty. They drink more and urinate more. Other reasons include drinking too much for reasons not related to thirst (in some animals this may be a form of stereotyped behavior) and over-hydration, or too much fluid given through other routes, such as intravenously. (See *pale urine*)

Dark urine

Description: An animal passes urine that is darker than normal for its sex, age and strain. This condition means that the waste products excreted by the body are more concentrated than normal.

Reasons: Dark urine may be seen if an animal is dehydrated or has kidney problems. Dark urine may also be normal for that species. This condition may be difficult to see in smaller animals. (See *too little urine*)

Too little urine (Oligouria)

Description: An animal has a decreased output of urine.

Reasons: Animals may make too little urine because they are dehydrated. They may also void smaller amounts of urine if they have a urinary tract infection or stones in the bladder or kidneys. If the urethra, the tube leading from the bladder to the outside, is blocked, animals will void no urine, quickly sicken and may die. It is normal for some species of desert rodents to make little urine. (See *dark urine, straining to urinate*)

Figure 52A Normal rodent feces. Clockwise from top left: Mice, hamsters, guinea pigs, rats, and gerbils.

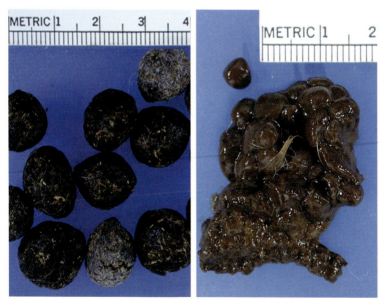

Figure 52B Normal feces from a rabbit. The harder, formed feces are day feces, and the less formed feces are night feces, which the rabbit normally ingests.

Figure 53 Loose feces in a mouse. Mice and rats rarely have watery diarrhea as other animals do. Instead, they have loose feces that stick to the perianal area. (Photo courtesy of LAVA, UK; www.digires.co.uk.)

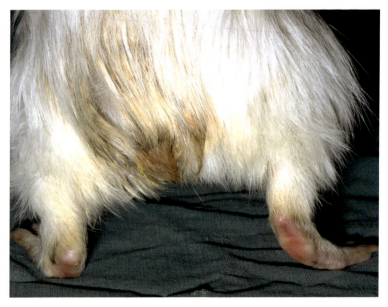

Figure 54A Diarrhea in a guinea pig. The feces are not shown, but the extensive fur staining indicates a serious problem. (Photo courtesy of LAVA, UK; www.digires.co.uk.)

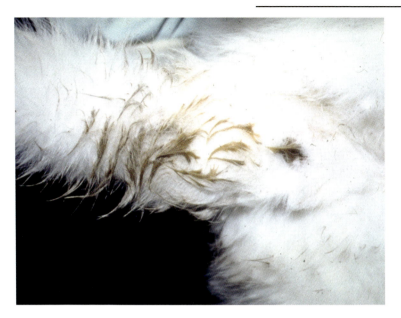

Figure 54B Diarrhea in a rabbit. Liquid feces are adherent to the fur of the underside of the tail. (Photo courtesy of LAVA, UK; www.digires.co.uk.)

Figure 54C Diarrhea in a hamster, also known as "wet tail." This is seen with a serious bacterial infection in hamsters.

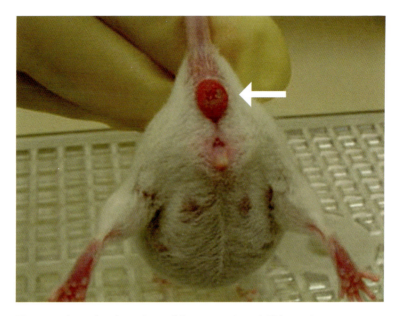

Figure 55 A rectal prolapse in an albino mouse (arrow). This may be seen secondary to certain bacterial infections or after giving birth.

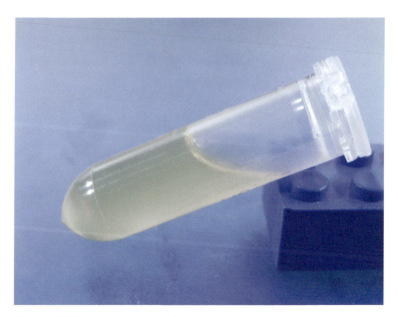

Figure 56A Normal mouse or rat urine.

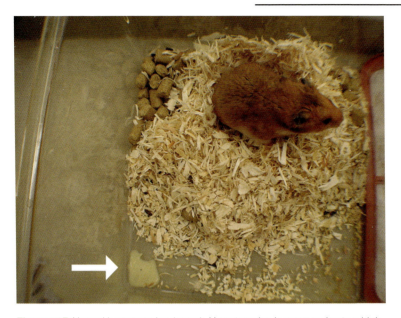

Figure 56B Normal hamster urine (arrow). Hamster urine is opaque due to a high level of minerals and protein.

Figure 56B Normal rabbit urine, showing some of the various colors and clarities possible with normal rabbits. The color and opacity is due to high levels of minerals and protein. (Photo courtesy of LAVA, UK; www.digires.co.uk.)

Reproductive System

Preputial or clitoral swelling
Figure 57

Description: Preputial swellings usually appear as one or two small bumps under the skin of the abdomen near the prepuce. The swellings are firm and may be red, or there may be pus coming from the lump or the prepuce.

Reasons: This condition is only found in males and almost entirely in mice. The prepuce is the skin covering the penis. Female mice have similar glands of the clitoris that may also become infected. Some books call these clitoral glands "preputial glands" in females. The glands of the prepuce are susceptible to infection and abscesses. This swelling of the prepuce and the glands above it is most often due to a bacterial infection.

Scrotal swelling
Figure 58

Description: This condition is found only in male animals. Scrotal swellings appear either as lumps on the scrotum or as an enlarged, rounded scrotum.

Reasons: Scrotal swelling may be due to trauma (injury) or herniation of abdominal contents (usually the intestines) into the scrotum. This condition will raise the temperature of the testicles and negatively affect breeding performance.

Penis stuck out of the prepuce (Paraphimosis, prolapsed penis)
Figure 59

Description: This condition is found only in male animals. An animal's penis is protruding from the prepuce and will not go back in.

Reasons: Trauma to the penis may result in injury and swelling that prevents it from entering the prepuce. Mouse urologic syndrome (MUS) may also result in paraphimosis.

Penis stuck in the prepuce (Phimosis)
Figure 60

Description: This condition is found only in male animals. An animal's penis cannot protrude from the prepuce.

Reasons: Phimosis is usually due to trauma to the opening of the prepuce, which prevents the penis from exiting. This condition can be secondary to preputial gland infections or injury. Urine dribbling may be seen with phimosis.

Vaginal or uterine prolapse
Figure 61

Description: This condition is seen only in females. An animal has pink or red tissue protruding from the vaginal area.

Reasons: This condition may occur after the animal gives birth. If animals strain to give birth to pups, occasionally ligaments rupture or stretch, allowing them to push the vagina or uterus outside the body.

Vaginal septum
Figure 62

Description: This condition is seen only in females, most often in mice and rats. It is a vertical flap of tissue in the vagina.

Reasons: This condition is a birth defect. The uterus of rodents and rabbits has two horns, one for each ovary. In the female fetus, the two horns of the uterus fuse into one vagina. In some animals, this fusion is not complete, and they are left with tissue dividing their vaginas as well as their uterus, making them unsuitable for breeding.

Vaginal closure membrane
Figure 63

Description: This condition is seen only in females. In rodents, before puberty, the vagina is closed from the outside by a membrane. In some cases, in rats and mice, this membrane will fail to dissolve at puberty. These animals may appear male as the secretions of the uterus and vagina will fill the vagina, giving them the appearance of a scrotum (female mice and rats have nipples, though, while males do not).

Reasons: Vaginal closure membranes are normal for guinea pigs unless they are receptive to mating. In other animals, it is not normal to keep a vaginal closure membrane after puberty, but the cause for retaining it is unknown and most probably genetic.

Difficulty giving birth (Dystocia, stuck pup)
Figure 64

Description: This condition is seen only in females. An animal has a pup stuck in the uterus, vagina, or part of the way out.

Reasons: Dystocia happens when the pup is too big for the mother to push out easily. It may also happen if the mother becomes too tired during the birth or if the pup is positioned incorrectly.

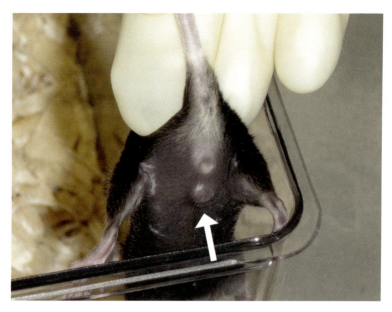

Figure 57A A preputial gland swelling in a C57BL/6 mouse. This swelling was an abscess.

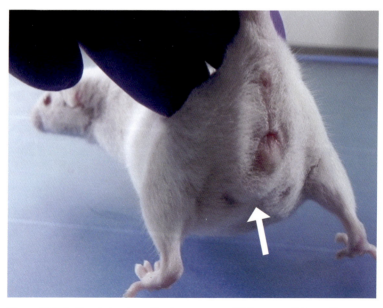

Figure 57B A clitoral gland swelling in a pregnant CD1 mouse. This swelling was also caused by an abscess.

Figure 58A Normal testicles in an albino mouse.

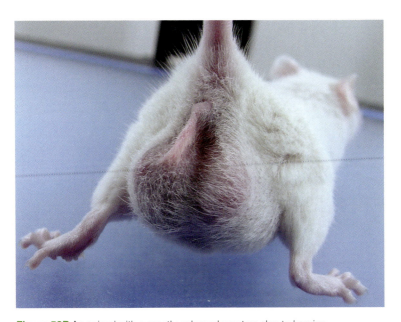

Figure 58B An animal with a greatly enlarged scrotum due to hernias.

Figure 58C Normal testicles in a nude mouse.

Figure 58D A nude mouse with an enlargement of one half of the scrotum (arrow) due to an inguinal hernia.

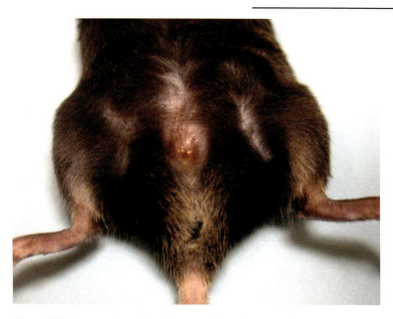

Figure 59 This mouse has phimosis, a condition when the penis is stuck inside the prepuce. This makes mating impossible and urination difficult. The problems in urination associated with phimosis often result in illness in the animal.

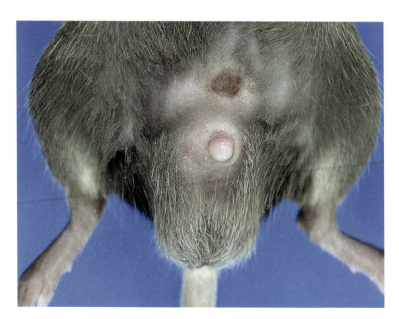

Figure 60 This mouse has paraphimosis, a condition where the penis is stuck outside the prepuce. The delicate tissues of the penis are easily damaged, resulting in an inability to mate and often difficulty urinating. This mouse also has a scab where a preputial gland abscess has ruptured to the surface.

Figure 61 This mouse has a vaginal prolapse and fur staining.

Figure 62A A normal vaginal opening in a C57BL/6 mouse. (Photo courtesy of Dr. Sonya Gearhart and Dr. Jennifer Kalishman)

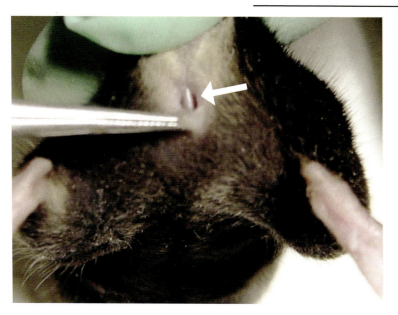

Figure 62B This mouse has a vaginal septum (arrow). The vertical line of tissue may impede mating and will certainly cause problems if the animal attempts to give birth. (Photo courtesy of Dr. Sonya Gearhart and Dr. Jennifer Kalishman)

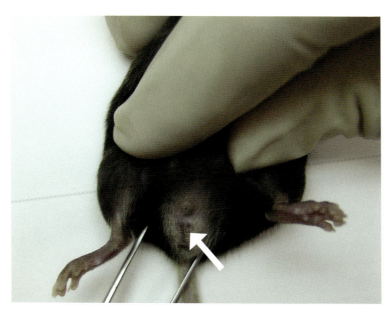

Figure 63 In this C57BL/6 mouse, the vaginal closure membrane has failed to dissolve. The arrow points to where an opening would normally be present (See Figure 62A).

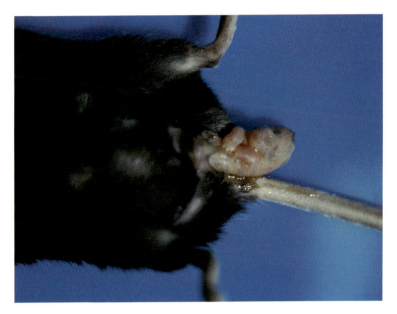

Figure 64 This animal has a dystocia or a difficult birth. The pup being born was stuck in the birth canal and the mouse became very ill.

Limbs and Paws

Missing toes
Figure 65

Description: An animal has fewer toes than normal.

Reasons: Sometimes, this condition is due to damage to the foot that happens in the cage when animals get feet stuck in cage bars or lids. If an animal severely damages a toe, it may fall off on its own. Missing toes may also be due to birth defects or genetic mutations. Sometimes, a mother will chew the toes of her pups.

Too many toes (polydactyly)

Description: An animal has more toes than normal.

Reasons: This condition is due to a spontaneous mutation.

Normal Number of Toes for Laboratory Rodents and Rabbits

Animal	Fore	Hind
Mouse	4	5
Rat	4	5
Gerbil	4	5
Hamster	4	4
Guinea pig	4	3
Rabbit	5	4

Broken limbs

Description: The bone of an animal's forelimb or hindlimb is broken.

Reasons: Broken limbs may be seen if animals are handled incorrectly, dropped or damaged by caging. Animals can break limbs on wire-bottomed cages (if they are not familiar with wire, they may catch a foot in it) or on wire cage lids.

Foot injuries
Figure 66

Description: Any injury that occurs to an animal's fore or hind foot.

Reasons: Foot injuries may be due to entrapment with fibers from nesting material. Females may chew the toes of pups. Animals may catch their feet or toes in wire or punched-bottomed cages.

Swollen joints (arthritis)
Figure 67

Description: An animal has joints that are larger than normal for its age and sex.

Reasons: Animals may have swollen joints due to injury, infectious disease or normal processes such as aging. Some swollen joints may be induced by investigators to study arthritis or other joint changes. Joint swelling may be seen on other parts of the body, such as the tail, but this is less common.

Foot calluses or foot sores (Pododermatitis)
Figure 68

Description: Calluses are a thickening of the skin on the bottom of the foot. Sores are open, ulcerated areas on the bottom of the foot.

Reasons: Animals may have calluses due to the cage in which they are housed. Animals housed on wire-bottomed cages may develop calluses on their paws. Calluses and sores may also develop on animals housed on bedding. Rarely, calluses may develop into sores. These foot conditions are more common in older or heavier animals. Animals may have foot sores due to trauma or fighting with other animals, as well. (See *ulceration*)

Stained forepaws
Figure 69

Description: An animal has matted, stained hair on the forepaws or forearms.

Reasons: Animals usually have stained forearms or forepaws when they have nasal or ocular discharge. The forepaws become stained from wiping the nose and eyes. (See *nasal discharge, ocular discharge, stained fur*)

Figure 65 This rat is missing most of the toes on both of its front feet. This may be due to damage from a cage, injury by its mother, or damage due to fighting. Genetic problems are also a possible cause of missing toes, but this is relatively rare.

Figure 66A This mouse has a swollen foot and leg. It is also hunched and ruffled.

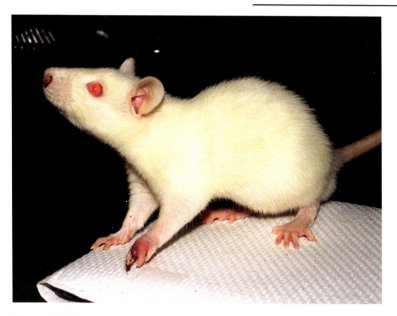

Figure 66B This rat has sustained an injury to the paw, resulting in the loss of a digit and generalized swelling of the paw. Injuries of this type may sometimes be seen with entrapment in wire cage lids.

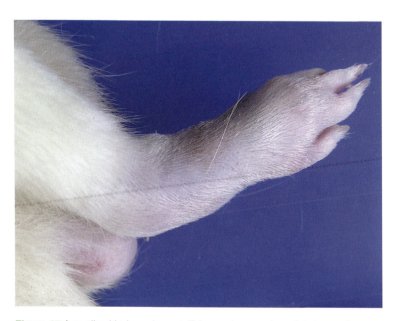

Figure 67 A swollen hind paw in a rat. This may be seen in certain types of research, typically on arthritis.

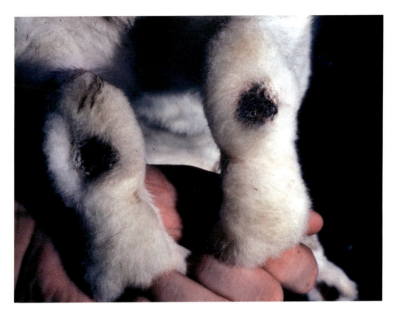

Figure 68A A condition known as "sore hocks" in a rabbit. This is sometimes seen in elderly, heavy rabbits kept on wire-bottomed cages. This is an extreme example.

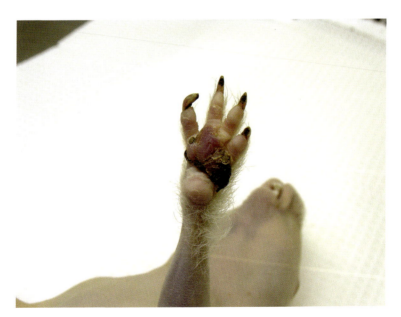

Figure 68B Foot calluses in a hairless guinea pig. Calluses may develop into open sores if the protective callus is disrupted (bacterial infection, sustained wetness, etc.)

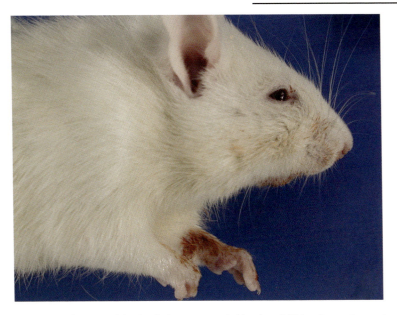

Figure 69A Staining of the forelimbs, eyes, and chin of a rat. This rat was stressed or ill, and the cleaning of the Harderian gland secretions from the eyes and nose resulted in these reddish stains on the paws and face.

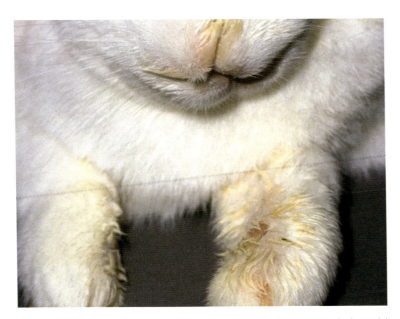

Figure 69B Staining of the paws and forelimbs of a rabbit. This rabbit had a bacterial infection of the nose and was wiping the nasal discharge away with its front feet.
(Photo courtesy of LAVA, UK; www.digires.co.uk.)

Ringtail
Figure 70

Description: An animal has circular constrictions of the skin on the tail.

Reasons: Ringtail is thought to be associated with humidity variations in the animal room, but the exact cause is not known. This condition can be mild or severe, resulting in loss of the tail, and is seen in mice and rats.

Tail injuries
Figure 71

Description: An animal has bites, scratches or ulcers on the tail. Injuries may also be caused by research interventions such as injections.

Reasons: Tail wounds are usually caused by animals fighting. They may also be caused by improper handling when using forceps. Tail damage may be seen as a consequence of damage to blood vessels in the tail. This condition may result in the death of the tip of the tail. Tails may become damaged due to fighting, blood collection, wire lids, or nesting material fibers entrapping the tail.

Kinked tail (Curly tail)
Figure 72

Description: An animal's tail has a kink, curl, or bend in it.

Reasons: Kinks or curls in the tail may be a genetic defect. They may also be caused by the healing of a previous break in the tail.

Short tail
Figure 73

Description: An animal's tail is shorter than normal for the age, sex and strain of the animal.

Reasons: Tails can be shortened due to amputations (removal of tail tips) for DNA sampling. Tails may be short due to damage of the tail by the cage. After the damage occurs, the dead tip of the tail falls off as part of the healing process. These tails tend to have blunt tips. Tails may also be short due to a mutation or defect in development.

Tail slip
Figure 74

Description: The skin of the tail of an animal slips off, leaving the muscles and tendons exposed.

Reasons: Tail slip is caused by improper handling of animals. Animals should always be handled as close to the base of the tail as possible, especially gerbils and heavy animals.

Figure 70 Ringtail of varying levels of severity in mice. The exact cause of ringtail is unknown, but the condition is linked to extremes of humidity.

Figure 71A This mouse has an injured tail (and hind feet) due to fighting. The tail is injured seriously enough that the blood supply to the tip of the tail has been cut off and that tissue has died.

Figure 71B Tail injuries due to fighting.

Figure 72A A kink in the tail of an albino mouse. Tail kinks like this can be genetic or can be due to improper handling resulting in broken tails.

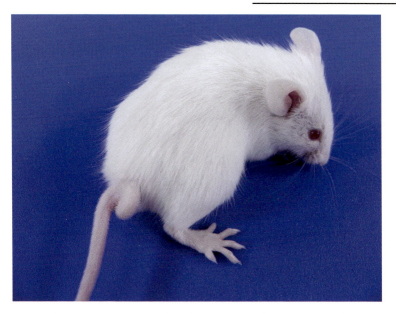

Figure 72B This mouse has a curled tail. This type of tail change is most likely due to a genetic mutation or a defect in development.

Figure 73A This animal has a short tail that also appears to be set in an odd position. A short tail like this is almost always due to a developmental defect.

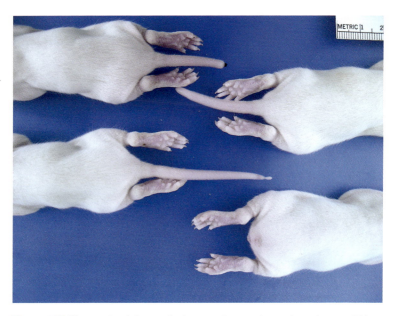

Figure 73B These animals have tails that are shortened to various degrees. This type of short tail is usually due to trauma — either injury by the mother or secondary to severe ringtail.

Figure 74 This photo of the tip of a gerbil's tail illustrates tail slip, where the skin of the tail has come off, exposing the tissue and bone. Gerbils are prone to this condition, but all rodents may slip their tails. Rodents should always be picked up by the base, not the tip, of the tail.

Commonly Used Medical Terms

1. HYPER- = Refers to an observation that is increased from normal.
2. LESION = a localized change or abnormality
3. MORBIDITY = illness or disease in a tissue or organ
4. MORTALITY = death
5. HYPO- = Refers to an observation that is decreased from normal.

Appendix B

NORMAL PHYSIOLOGIC VALUES

	Mouse	Rat	Hamster	Guinea Pig	Gerbil	Rabbit
Body temperature	35.8-37.6°C 96.4-99.7°F	35.9-37.5°C 96.6-99.5°F	37-38°C 98.6-100.4°F	37-39.5°C 98.6-103.1°F	35.7-39.3°C 96.3-102.7°F	38-40°C 100.4-104°F
Heart rate (beats per minute)	328-780	250-600	250-600	230-320	260-600	130-325
Respiration rate (per minute)	90-220	66-114	35-120	42-104	70-120	30-60
Weight range (g): adult male	25-40	300-500	80-120	500-800	80-90	2-6 kg
Weight range (g): adult female	25-40	200-400	80-120	500-800	70-80	2-6 kg
Weight: neonate (g)	1	5	8-12	70-90	3	30-80
Water consumption (daily; ml)	4-7	24-60	10	50-80	3-6	100-600
Food consumption (daily; g)	3-6	15-30	10	30-48	4-7	100-300
Life span (years)	1-3	2.5-3.5	1.5-2	4-6	2-4	5-8
Age at sexual maturity	40-60 d	65-110 d	42-70 d	60-90 d	70-84 d	4-6 m
Estrous cycle frequency	4-5 d	4-5 d	4 d	16-18 d	24 h	Induced
Duration of estrus (h)	10	13-15	20	6-11	24	NA
Gestation period (d)	19-21	20-22	15-16	60-65	24-26	29-35
Average litter size	6-10	7-12	5-10	2-4	4-5	4-10
Nursing frequency (per day)	>10	>10	>10	>10	>10	~1
Young begin eating dry food (d)	10-12	10-12	7-10	4-6	16-20	21
Age at weaning (d)	21-28	21	21	7-21	21	28-42
Breeding life	8 m	1.5 y	1 y	2-4 y	3 y	3 y

g=grams   ml=milliliters   h= hours   d=days   m=months   y=years